Methods of Nonlinear Analysis

Methods of Nonlinear Analysis

Editor

Alex Wilson

scitus academics

Methods of Nonlinear Analysis
Edited by **Alex Wilson**

ISBN: 978-1-68117-645-1
Library of Congress Control Number: 2016946506

www.scitusacademics.com

Notice

Printed in United States of America on Acid Free Paper

Preface

All physical processes are inherently nonlinear to a certain extent. For example, when you stretch a rubber band, it gets harder to pull as the deflection increases; or when you flex a paper clip, permanent deformation is achieved. Several common every day applications like these exhibit either large deformations and/or inelastic material behavior. Failure to account for nonlinear behavior can lead to product failures, safety issues, and unnecessary cost to product manufacturers. Over the last decade, finite element analysis (FEA) stopped being regarded only as an analyst's tool and entered the practical world of design engineering. CAD software now comes with built-in FEA capabilities and design engineers use FEA as an everyday design tool in support of the product design process. However, until recently, most FEA applications undertaken by design engineers were limited to linear analysis. Such linear analysis provides an acceptable approximation of real-life characteristics for most problems design engineers encounter. Nevertheless, occasionally more challenging problems arise, problems that call for a nonlinear approach. Historically, engineers were reluctant to use nonlinear analysis, because of its complex problem formulation and long solution time. That's changing now, as nonlinear FEA software interfaces with CAD and has become much easier to use. In addition, improved solution algorithms and powerful desktop computers have shortened solution times. A decade ago, engineers recognized FEA as a valuable design tool. Now they are starting to realize the benefits and greater understanding that nonlinear FEA brings to the design process.

Methods of Nonlinear Analysis is devoted to nonlinear problems coming from different areas, with particular reference to those introducing new techniques capable of solving a wide range of problems. The book focuses on significant problems in pure and applied nonlinear analysis. It seeks to present the most significant advances in this field to a wide readership, including researchers and graduate students in mathematics, physics, and engineering.

TABLE OF CONTENTS

1
CHAPTER

Singular Solutions for a Class of Traveling Wave Equations Arising in Hydrodynamics

Anna Geyer[a,*], **Víctor Mañosa**[b]

[a] *Faculty of Mathematics, University of Vienna, Oskar-Morgenstern-Platz 1, 1090 Vienna, Austria*

[b] *Departament de Matemàtiques, Control, Dynamics and Applications Group (CoDALab), Universitat Politècnica de Catalunya, Colom 1, 08222 Terrassa, Spain*

ABSTRACT

We give an exhaustive characterization of singular weak solutions for ordinary differential equations of the form $\ddot{u}\,u + \frac{1}{2}\dot{u}^2 + F'(u) = 0$, where F is an analytic function. Our motivation stems from the fact that in the context of hydrodynamics several prominent equations are reducible to an equation of this form upon passing to a moving frame. We construct peaked and cusped waves, fronts with finite-time decay and compact solitary waves. We prove that one cannot obtain peaked and compactly supported traveling waves for the same equation. In particular, a peaked traveling wave cannot have compact support and vice versa. To exemplify the approach we apply our results to the Camassa–Holm equation and the equation for surface waves of moderate amplitude, and show how the different types of singular solutions can be obtained varying the energy level of the corresponding planar Hamiltonian systems.

Keywords: Camassa–Holm equation; Integrable vector fields; Singular ordinary differential equations; Traveling waves

1. INTRODUCTION

In the present paper we propose to study certain types of weak solutions for ordinary differential equations (ODE) of the form

$$\ddot{u}\,u + \frac{1}{2}\dot{u}^2 + F'(u) = 0, \tag{1}$$

where F is an analytic function. Our motivation stems from the fact that a variety of model equations arising in the context of hydrodynamics, among them the well-known Camassa–Holm equation (cf. [1], [2] and [3]) and the related equation for surface waves of moderate amplitude (cf. [4], [5], [6], [7] and [8]), are reducible to an ODE of the form (1) upon passing to a moving frame. Owing to the fact that every solution of Eq. (1) may be interpreted as a traveling wave of a suitable underlying partial differential equation (PDE) we will call the solutions of (1) traveling waves.

The singular nature of Eq. (1) accounts for the non-uniqueness of certain solutions, which we call *singular solutions*. These are in general weak solutions, but have stronger regularity than one would expect a priori: the solutions are analytic except for a countable number of points at which the equation is satisfied in the limit. Furthermore, Eq. (1) admits an order reduction which allows us to see that under certain conditions on F, the solutions are actually classical solutions of this reduced equation.

The main result of this paper consists in the exhaustive characterization of singular solutions of (1) from qualitative properties of the function F(u). We show that Eq. (1)admits solutions with peaks and cusps, fronts with finite-time decay and solitary solutions with compact support. Furthermore, we find that one cannot obtain peaked and compactly supported solutions for the same F. In particular, a peaked solution cannot have compact support and vice versa. The characterization of classical solutions of (1)will be covered only very briefly for the convenience of the reader, since our main focus lies in the analysis of singular solutions.

We apply our results to the aforementioned nonlinear partial differential equations, and show how the different types of singular solutions are obtained varying the energy levels of the Hamiltonian planar differential system corresponding to (1). It lies beyond the scope of this paper to prove in full generality that every weak solution of (1) is also a weak traveling wave solution of an underlying PDE. For a discussion of this problem we refer the reader to [9] and [10], where it is shown that in the special case of the Camassa–Holm equation every weak solution of (1) is a weak traveling wave solution of the underlying PDE. Following similar steps the same result can be shown for the equation of surface waves of moderate amplitude.

The structure of the paper is as follows. In Section 2, we give the precise definitions of weak and singular solutions and provide a preliminary result on the non-uniqueness of solutions of (1). In Section 3 we introduce the notion of *elementary forms*, classical solutions of (1) defined on a subset of R, from which we construct singular solutions. Furthermore, we discuss how the qualitative features of any traveling wave solution can be obtained from the properties of F. The main results of the paper, Proposition 6, Proposition 7, Proposition 11 and Proposition 12 are presented in Section 4 which is devoted to the complete characterization of singular solutions. In Section 5, we characterize the classical and singular traveling waves of the Camassa–Holm equation and the equation for surface waves of moderate amplitude in shallow water.

2. WEAK AND SINGULAR SOLUTIONS

Our focus lies in the characterization of solutions which are not classical, so we require a weak formulation of (1). Keeping in mind that any solution of (1) can be interpreted as a traveling wave of an underlying PDE, we will consider only bounded solutions.

Definition 1.

We say that a bounded function $u \in H^1_{loc}(\mathbb{R})$ is a *traveling wave solution (TWS)* if it satisfies (1) in the sense of distributions, i.e. if u satisfies

$$\int_{\mathbb{R}} (u^2)_t \phi_t + (u_t)^2 \phi - 2F'(u)\phi \, dt = 0, \tag{2}$$

for any test function $\phi \in C_c^\infty(\mathbb{R})$. We say that u is a *strong TWS* if it satisfies (1) in the classical sense.

It turns out that the concept of weak solutions is quite crude. Indeed, if no further conditions are imposed it is possible to find a plethora of weak solutions of (1) giving rise to TWS with very complex shapes. For instance it is known that the Camassa–Holm equation can have TWS of the form $u=\varphi(t)$ such that some of its level sets $\{\varphi(t)=k\}$ are cantor sets, cf. [10]. In the present paper, however, we are interested in those solutions which fail to be strong TWS because of the singularity, but which still have a certain degree of regularity, and in fact are strong solutions of (1) except for a finite or a countable set of points (such solutions are, for instance, peaked or cusped waves, fronts with finite time decay, compact solitary waves and composite waves, see Section 4 for precise definitions). To see this, observe that Eq. (1) admits the order reduction

$$\frac{d}{dt}\Big(\frac{u\dot{u}^2}{2}+F(u)\Big)=\dot{u}\left(\ddot{u}\,u+\frac{1}{2}\dot{u}^2+F'(u)\right),$$

which is equivalent to the fact that the planar differential system associated to (1) has a first integral. Therefore, any classical solution of (1) naturally satisfies

$$\frac{u\dot{u}^2}{2}+F(u)=h, \quad \text{for some constant } h\in\mathbb{R}. \tag{3}$$

The singularity of (3) leads to the existence of non constant solutions $u\in H^1_{loc}(\mathbb{R})$ of (2)which satisfy (3) except, perhaps, at a countable number of points where the derivative is not defined but the equation is still satisfied in the limit. This motivates the following definition:

Definition 2.

Let $u(t)\in H^1_{loc}(\mathbb{R})$ be a non constant TWS of (1). Then u (t) is called a

(a)*strong singular* TWS of (1) if u is a classical solution of (3) on R.

(b)*weak singular* TWS of (1) if u is a classical solution of (3) on $R\setminus S$, where S is the set of countably many points t_k such that

$$\lim_{t\to t_k} \frac{u(t)\dot{u}(t)^2}{2} + F(u(t)) = h. \tag{4}$$

For the remainder of this paper we set $h_0 := F(0)$. The next result establishes the non-uniqueness of solutions of (3) for $h = h_0$. This will play a role in the construction of singular TWS of (1), see Section 4.

Lemma 3.

Consider Eq. (3) with F *an analytic function such that* $u(h-F(u))>0$ *for* $u\in U\setminus\{0\}$ *where* U *is a neighborhood of* 0.

(a)*If* $h\neq h_0$, *then Eq.* (3) *is not defined at* $u=0$.

(b)*If* $h=h_0$ *then* $u\equiv 0$ *is a solution of Eq.* (3). *Furthermore if* $F'(0)\neq 0$ *or* $F'(0)=0$ *and* $F''(0)\neq 0$ *then Eq.* (3) *is not Lipschitz continuous in* $u=0$.

Proof.

Notice that Eq. (3) may be written as

$$\dot{u} = \pm\sqrt{2\frac{h-F(u)}{u}} =: v_h^{\pm}(u).$$

Observe that if $h\neq h_0$ then $\lim_{u\to 0} v_h^{\pm}(u) = \pm\infty$, which proves statement (a). When $h=h_0$, $u\equiv 0$ is always a solution of (3). Note that

$$\frac{|v_{h_0}^{\pm}(u)|}{|u|} = \sqrt{-2\left(\frac{F'(0)}{u^2} + \frac{F''(0)}{2!\,u} + \frac{F^{(3)}(0)}{3!} + o(u)\right)}.$$

Hence,

if $F'(0)\neq 0$ or $F'(0)=0$ and $F''(0)\neq 0$ then $\lim_{u\to 0} |v_{h_0}^{\pm}(u)|/|u| = +\infty$.
Therefore it is not possible to find a constant $L>0$ such

that $|v_{h_0}^{\pm}(u)| \leq L|u|$ for $u \in U$, and hence the right-hand side of the differential equation fails to be Lipschitz continuous in $u=0$.

Remark 4.

Observe that $v_{h_0}^{\pm}(u)$ is Lipschitz continuous in $u=0$ only if $F'(0)=F''(0)=0$ and $F^{(3)}(0)\neq 0$. However, as a consequence of Proposition 6, Proposition 7 and Proposition 11, such F do not yield any singular solutions, and therefore they will not be considered in this paper.

3. TRAVELING WAVE SOLUTIONS FROM QUALITATIVE PROPERTIES OF F

The planar system associated to Eq. (1) is given by

$$\begin{cases} \dot{u} = v \\ u\,\dot{v} = -F'(u) - \dfrac{1}{2}\,v^2. \end{cases} \tag{5}$$

A straightforward computation shows that system (5) possesses the first integral

$$H(u,v) = \frac{u\,v^2}{2} + F(u), \tag{6}$$

whose *energy* $H(u,v)=h$ is therefore constant along solutions. We will study how the TWS of (1) can be obtained from the orbits of the associated system (5), and how these orbits depend on the qualitative properties of F. Our study resembles the study of conservative systems of one degree of freedom [11, Section 12]. For such systems, the qualitative features of the phase portrait are in correspondence with the ones of the potential function. Analogously, we find that the qualitative features of the function F characterize the phase portrait of (5) and therefore the types of TWS that (1) may exhibit. In contrast to potential systems, the Hamiltonian (6) has no purely kinetic term, which leads to the presence of a singular line in the phase portrait and the existence of solutions defined on subsets of R. Without loss of generality we consider the system only in the half-plane $(\{u>0\}\cup\{u=0\})\times R$, which is not restrictive since our analysis is of a qualitative nature. It is worth mentioning however that in some applications it might be interesting to consider solutions in $\{u\leq 0\}$.

The solutions of Eq. (1) are associated to the orbits of (5), which correspond to the level sets of the energy $H(u,v)=h$, i.e. they lie on curves of the form $\gamma_h=\{H=h\}$ which are composed of two symmetric branches $(u, v_h^{\pm}(u))$, where

$$v_h^{\pm}(u) = \pm\sqrt{2\,\frac{h-F(u)}{u}}. \tag{7}$$

Therefore, solutions of Eq. (1) corresponding to energy h exist if there is a non-empty set $I\subseteq\{u\geq 0\}$ such that $F(u)-h<0$ for all $u\in I$. Recall that our interest lies in the analysis of bounded solutions, which correspond to branches of the curves γ_h that are bounded in the u-direction. These branches are defined if and only if either there exist $m_1,m_2>0$ such that $F(m_1)=F(m_2)=h$ and $F(u)<h$ for all $u\in(m_1,m_2)$, or there exists $m>0$ such that $F(m)=h$ and $F(u)<h$ for all $u\in(0,m)$. The latter case corresponds to a type of strong solutions of (1) whose maximal domain of definition is a (finite) subset of R. We call these solutions *elementary forms*, which play an important role in our construction of singular TWS. The former case corresponds to classical smooth solutions which are uniquely defined on R. Observe that possible oscillations of F in (m_1,m_2) or $(0,m)$ affect the solutions of (1) only at the level of their convexity.

3.1. Smooth TWS on R

In this subsection we briefly summarize how smooth TWS are characterized in terms of qualitative aspects of F. Assume that there exist $m_1,m_2>0$ such that $F(m_1)=F(m_2)=h_m$ and $F(u)<h_m$ for

all $u \in (m_1, m_2)$. Then the two branches of γ_{hm} given by $(u, v_{h_m}^{\pm}(u))$ are defined for $u \in [m_1, m_2]$ and coincide at the points $p_1 = (m_1, 0)$ and $p_2 = (m_2, 0)$.

Observe that the critical points of system (5) in $\{u > 0\} \times R$ are of the form $(u, 0)$ where $F'(u) = 0$. Hence a point $p_i = (m_i, 0)$ will be a critical point if $F'(m_i) = 0$ and a regular point otherwise. We distinguish between the following cases (see Fig. 1):

(A) If $F'(m_i) \neq 0$ for $i = 1, 2$, then the points p_1 and p_2 are regular, and the two branches $(u, v_{h_m}^{\pm}(u))$ give rise to an isolated closed curve, which is a periodic orbit of (5). This periodic orbit corresponds to a strong smooth periodic solution $u(t)$ of Eq. (1). Observe that since $v_{h_m}^{+}(u)$ and $v_{h_m}^{-}(u)$ are not multivalued functions, the closed curve has no lobes and therefore $u(t)$ reaches a unique local minimum at m_1 and a unique local maximum at m_2 in each period. Furthermore, this periodic solution is symmetric with respect to its local minima and maxima.

(B) If $F'(m_1) = 0$ and $F'(m_2) \neq 0$ then p_1 is a critical point and p_2 is a regular one. In this case, the branches $(u, v_{h_m}^{\pm}(u))$ yield a homoclinic loop giving rise to a smooth solitary wave solution of (1) which has a unique maximum at m_2 and which decays exponentially on either side of the maximum such that $\lim_{t \to \pm\infty} u(t) = m_1$ (recall that since p_1 is a critical point, the time that the orbit of (5) takes to leave from or to reach this point is infinite). Furthermore, this solitary wave solution is symmetric with respect to its maximum. If $F'(m_1) \neq 0$ and $F'(m_2) = 0$ then there appears a solitary wave of (1) with a unique minimum at m_1 and such that $\lim_{t \to \pm\infty} u(t) = m_2$.

(C) If $F'(m_1) = F'(m_2) = 0$ then p_1 and p_2 are critical points, and the branches $(u, v_{h_m}^{\pm}(u))$ yield a heteroclinic loop connecting these points in an infinite time. This pair of connecting orbits gives rise to a smooth front decaying from m_2 to m_1 such that $\lim_{t \to -\infty} u(t) = m_2$ and $\lim_{t \to +\infty} u(t) = m_1$, and another smooth front increasing from m_1 to m_2 such that $\lim_{t \to -\infty} u(t) = m_1$ and $\lim_{t \to +\infty} u(t) = m_2$.

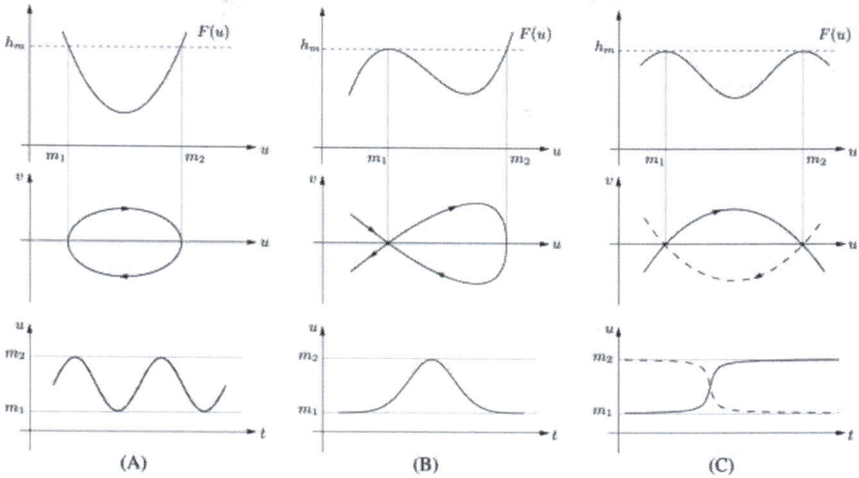

Figure 1. Global strong TWS of (1). (A) Periodic wave; (B) Solitary wave or pulse; (C) Smooth front.

3.2. Elementary forms: smooth TWS on a subset of R

In this subsection we discuss two types of elementary forms. The first type, studied in cases (a) and (b) below, are the ones associated to the curves γ_{h0} which correspond to the energy level $h_0 = F(0)$ and which intersect the singular line $\{u=0\} \times R$ at a finite point. The second type are the ones associated to curves γ_h that are unbounded in the component v and are arbitrarily close to the singular line $\{u=0\} \times R$ at infinity. They are studied in the case (c). In both cases, it will be a key step to study whether the orbits of (5)approach the singular line $u=0$ in finite time. To this end we prove the following auxiliary result.

Lemma 5.

Let F *be an analytic function. Suppose that there exist* $\varepsilon > 0$ *and* $m > 0$ *such that* $F(m) = F(0)$ *and* $F(u) < F(0)$ *for* $u \in (0, \varepsilon) \cup (m-\varepsilon, m)$.

(i)*Let* $n \in N$ *be the lowest order such that* $F^{(n)}(0) \neq 0$. *If* $n \leq 2$ *then*

$$\int_0^{\varepsilon} \sqrt{\frac{u}{F(0) - F(u)}} du \qquad (8)$$

is convergent. Otherwise the integral diverges.

(ii) *If* $F'(m)>0$, *then*

$$\int_{m-\varepsilon}^{m} \sqrt{\frac{u}{F(0)-F(u)}}du \qquad (9)$$

is convergent. Otherwise the integral diverges.

Proof.

In both cases we will apply the following criterion: let $f(x)$ be an unbounded function at $x=c$ such that $f(x)\geq 0$ and $\lim_{x\to c} f(x)|x-c|^{k}=A$, where $A\neq\infty$ and $A\neq 0$. Then, for $k<1$ the integral $\int_a^c f(x)dx$ is convergent, whereas for $k\geq 1$ it is divergent. To prove (i), set

$$f(u) := \sqrt{\frac{u}{F(0)-F(u)}},$$

and $n\in N$ such that $F^{(n)}(0)<0$ and $F^{(k)}(0)=0$ for $k<n$. Hence

$$f(u) = \sqrt{\frac{-n!}{F^{(n)}(0)u^{n-1}+o(u^{n})}},$$

and

$$\lim_{u\to 0+} f(u)|u|^{\frac{n-1}{2}} = \sqrt{\frac{-n!}{F^{(n)}(0)}} =: K, \quad \text{where } 0\neq K\neq\infty.$$

Therefore, using the criterion stated above, for $n\leq 2$ the integral (8) converges and otherwise it diverges. To prove (ii) let

$$f(u) := \sqrt{\frac{u}{F(m)-F(u)}},$$

and $n \geq 1$ such that $F^{(n)}(m)<0$ and $F^{(k)}(m)=0$ for $k<n$. Then

$$f(u)=\sqrt{\frac{-n!u}{F^{(n)}(m)(u-m)^n+o((u-m)^{n+1})}},$$

and therefore

$$\lim_{u\to 0+} f(u)|u-m|^{\frac{n}{2}}=\sqrt{\frac{-n!m}{F^{(n)}(m)}}=:K,\quad \text{where } 0\neq K\neq\infty.$$

Hence for $n=1$ the integral (9) converges and otherwise it diverges. ■

We start with a discussion of solutions associated to the energy level $h=h_0$. Suppose that there exists $m>0$ such that $F(m)=h_0$ and $F(u)<h_0$ for all $u\in(0,m)$. In this case, the curves γ_{h0} intersect the singular line $\{u=0\}$ at a finite point, and the branches $(u, v_{h_0}^{\pm}(u))$ are defined for $u\in[0,m]$, and coincide at the point $p=(m,0)$. As in the discussion for smooth solutions in Section 3.1 we consider the following cases:

(a)Suppose that $F'(0)\neq 0$. Then,

$$\lim_{u\to 0+} v_{h_0}^{\pm}(u)=\lim_{u\to 0+}\pm\sqrt{2\frac{F(0)-F(u)}{u}}=\pm\sqrt{-2F'(0)}=:\pm a\neq\pm\infty$$

and therefore the branches $(u, v_{h_0}^{\pm}(u))$ intersect the singular line $\{u=0\}$ at two distinct points $(0,\pm a)$. Moreover, they reach the singular line in finite time, sinceLemma 5(i) guarantees that the orbit γ_{h0} of system (5) connects the point $(0,+a)$ with the point $(\bar{u}, v_{h_0}^{+}(\bar{u}))$ in *finite time* $\Delta T(\bar{u})$. Indeed, by direct integration of system(5) with $v=v_{h_0}^{+}(u)$, and recalling that $F'(0)<0$, we have

$$\Delta T(\bar{u}) = \int_0^{\bar{u}} \frac{du}{v} = \int_0^{\bar{u}} \sqrt{\frac{u}{2(h_0 - F(u))}}\,du = \frac{1}{\sqrt{2}} \int_0^{\bar{u}} \sqrt{\frac{u}{F(0) - F(u)}}\,du < \infty. \quad (10)$$

(a1) If $F'(m) \neq 0$, then p is a regular point and there is an isolated orbit of (5)connecting the points $(0,+a)$ and $(0,-a)$, see Fig. 2(a1). By the above arguments, this orbit connects these points in a finite time, hence it gives rise to a strong solution $u(t)$ of (1) defined for some finite interval (t_1,t_2). As in the smooth cases, this isolated orbit defines a curve with no lobes and thereforeu (t) has a unique maximum m at $t=(t_1+t_2)/2$. This solution is symmetric with respect to its maximum due to the symmetry of the curve γ_{h0}.

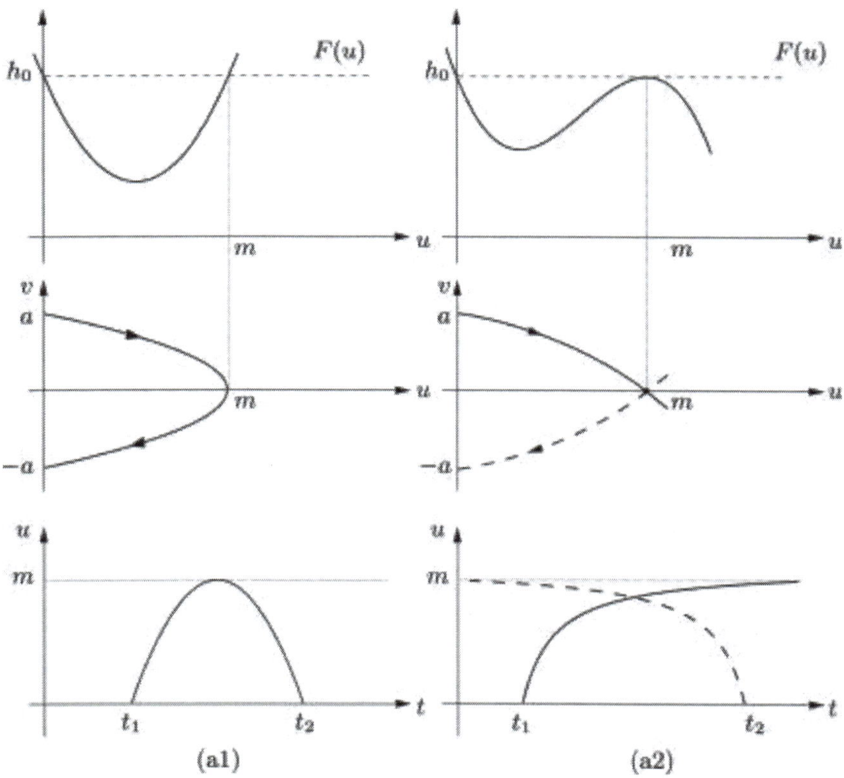

Figure 2. Strong solutions of (1) defined in a subset of R with energy $h=h_0$ when $F'(0)\neq 0$. These provide the elementary forms to construct peaked periodic and solitary TWS.

(a2) If $F'(m)=0$, then p is a critical point and the branch $(u, v_{h_0}^{+}(u))$ defines an orbit leaving its α-limit $(+a,0)$ in finite time and reaching its ω-limit p in infinite time. This orbit corresponds to a strong solution $u_+(t)$ of (1) defined for some interval$(t_1,+\infty)$, $t_1 \in R$, such that $\lim_{t\to t_1^+} u_+(t) = 0$, $\lim_{t\to t_1^+} \dot{u}_+(t) = a$ and $\lim_{t\to\infty} u_+(t)=m$, Fig. 2(a 2). Analogously, the branch $(u, v_{h_0}^{-}(u))$ gives an orbit such that each point on the orbit connects with the point p in an infinite time, and with$(-a,0)$ in a finite time. This gives a strong solution $u_-(t)$ of (1) defined for some interval $(-\infty,t_2)$, $t_2 \in R$, such that $\lim_{t\to t_2^-} u_-(t) = 0$, $\lim_{t\to t_2^-} \dot{u}_-(t) = -a$ and$\lim_{t\to-\infty} u_-(t)=m$. Notice that $u_+(t_1-t)=u_-(t_1-t)$ when $t_1=t_2$ due to the symmetry of system (5).

(b) Suppose that $F'(0)=0$. Then,

$$\lim_{u\to 0+} v_{h_0}^{\pm}(u) = \pm\sqrt{-2F'(0)} = 0.$$

Hence the branches $(u, v_{h_0}^{\pm}(u))$ are defined for $u \in [0,m]$, and coincide in $(0,0)$ and $p=(m,0)$.

(b1) If $F'(m) \neq 0$, then p is a regular point. Hence, there is an isolated orbit of (5) in$\{u>0\} \times R$ whose α and ω–limit is $(0,0)$, Fig. 3(b1). Lemma 5(i) and (ii) ensure that the orbit of (5) connects $(0,0)$ with $(m,0)$ in finite time if $F''(0)<0$ and $F'(m)>0$. Therefore, it corresponds to a strong solution $u(t)$ of (1) defined in a finite interval (t_1,t_2), such that $\lim_{t\to t_1^+} u(t) = \lim_{t\to t_2^-} u(t) = 0$ and $\lim_{t\to t_1^+} \dot{u}(t) = \lim_{t\to t_2^-} \dot{u}(t) = 0$. In addition, $u(t)$ reaches a unique local maximumm at $t=(t_1+t_2)/2$ and is symmetric with respect to this maximum due to the symmetry of system (5).

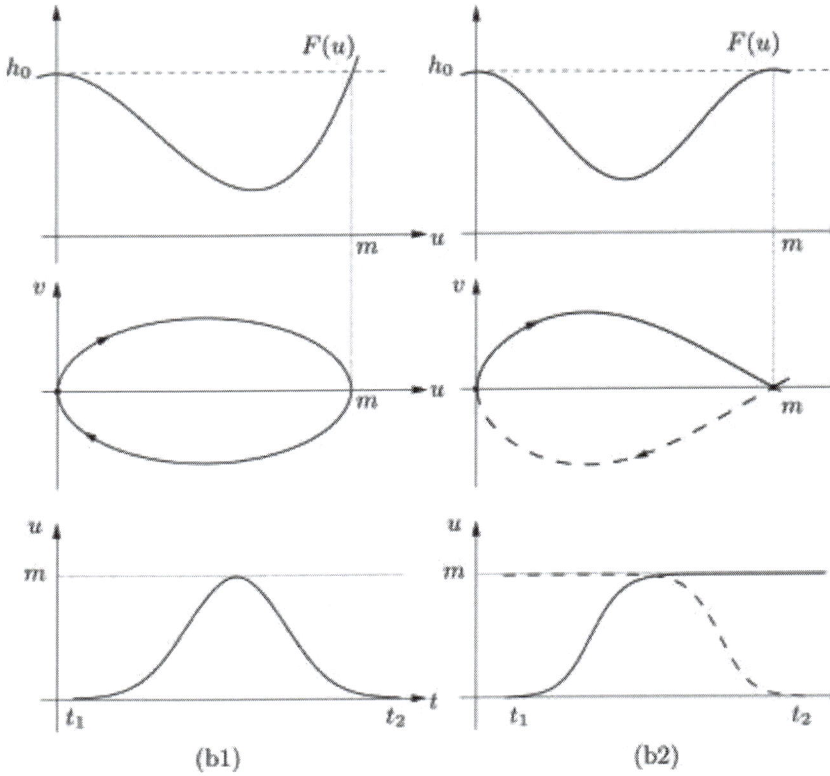

Figure 3. Strong solutions of (1) defined in a subset of R with energy $h=h_0$ when $F'(0)=0$. These provide elementary forms to construct TWS of class $C^1(R)$ (with compact support).

(b2) If $F'(m)=0$, then p is a critical point and there are two orbits with energy h_0 connecting $(0,0)$ with p, cf. Fig. 3(b2). In view of Lemma 5, any point of this orbit connects with $(0,0)$ in finite time, thus the heteroclinic orbit connecting $(0,0)$ with p through the branch $(u, v_{h_0}^+(u))$ corresponds to a strong solution $u_+(t)$ of (1) defined in (t_1,∞), $t_1 \in R$, such that $\lim_{t\to t_1^+} u_+(t) = \lim_{t\to t_1^+} \dot{u}_+(t) = 0$ and $\lim_{t\to\infty} u(t)=m$. The branch $(u, v_{h_0}^-(u))$ corresponds to a strong solution $u_-(t)$ defined in $(-\infty,t_2)$ with $t_2 \in R$ such that $\lim_{t\to t_2^-} u_-(t) = \lim_{t\to t_1^+} \dot{u}_-(t) = 0$ and $\lim_{t\to-\infty} u(t)=m$. Notice that $u_+(t_1+t)=u_-(t_1-t)$ when $t_1=t_2$ due to the symmetry of system (5).

The second type of solutions are the ones associated to curves γ_h which are defined for $u\in(0,m]$ but do not tend to a finite point on the singular line $\{u=0\}\times R$, hence they satisfy

$$\lim_{u\to 0+} v_h^{\pm} = \pm\infty.$$

Such curves appear if and only if there exists $m>0$ such that $F(u)<h_m=F(m)$ for all $u\in[0,m)$, see Fig. 4. In addition, observe that for $0<\bar{u}<m$

$$\Delta T(\bar{u}) = \int_0^{\bar{u}} \frac{du}{v} = \frac{1}{\sqrt{2}}\int_0^{\bar{u}} \sqrt{\frac{u}{F(m)-F(u)}}du < \infty, \tag{11}$$

since $F(m)-F(u)>0$ for $u\in[0,m)$. Consequently, the orbits of (5) passing through the points $(\bar{u}, v_{h_m}^{\pm}(\bar{u}))$ are unbounded in the component v but approach the singular line $u=0$ at infinity in finite time. Analogously to the previous cases, we distinguish between two scenarios:

(c1) If $F'(m)\neq 0$ then the branches $(u, v_{h_m}^{\pm}(u))$ coincide at the regular point $p=(m,0)$. Thus there is an unbounded orbit of (5) tending to infinity in the v-direction in finite time, Fig. 4(c1). It corresponds to a strong solution $u(t)$ of (1) defined in a finite interval (t_1,t_2) such that $\lim_{t\to t_1^+} u(t) = \lim_{t\to t_2^-} u(t) = 0$ and $\lim_{t\to t_1^+} \dot{u}(t) = +\infty$, $\lim_{t\to t_2^-} \dot{u}(t) = -\infty$. This solution has a unique maximum m at $t=(t_1+t_2)/2$ and it is symmetric with respect this point.

(c2) If $F'(m)=0$ then the branches $(u, v_{h_m}^{\pm}(u))$ coincide at the critical point $p=(m,0)$ and there are two unbounded orbits of (5) tending to infinity in the v-direction in finite time, cf. Fig. 4(c2), but any point in the orbit takes an infinite time to reach the singular point p. The orbit defined by the branch $(u, v_{h_m}^{+}(u))$ corresponds to a strong solution $u_+(t)$ of (1) defined for $t \in (t_1, \infty)$, $t_1 \in R$, such that $\lim_{t \to t_1^+} u_+(t) = 0$, $\lim_{t \to t_1^+} \dot{u}_+(t) = +\infty$, and $\lim_{t \to \infty} u_+(t) = m$. Similarly, the orbit defined by the branch $(u, v_{h_m}^{-}(u))$ corresponds to a strong solution $u_-(t)$ of (1) defined for $t \in (-\infty, t_2)$, $t_2 \in R$, such that $\lim_{t \to -\infty} u_-(t) = m$ and $\lim_{t \to t_2^-} u_-(t) = 0$, $\lim_{t \to t_2^-} \dot{u}_-(t) = -\infty$. Notice that as before, $u_+(t_1 + t) = u_-(t_1 - t)$ when $t_1 = t_2$ due to the symmetry of system(5).

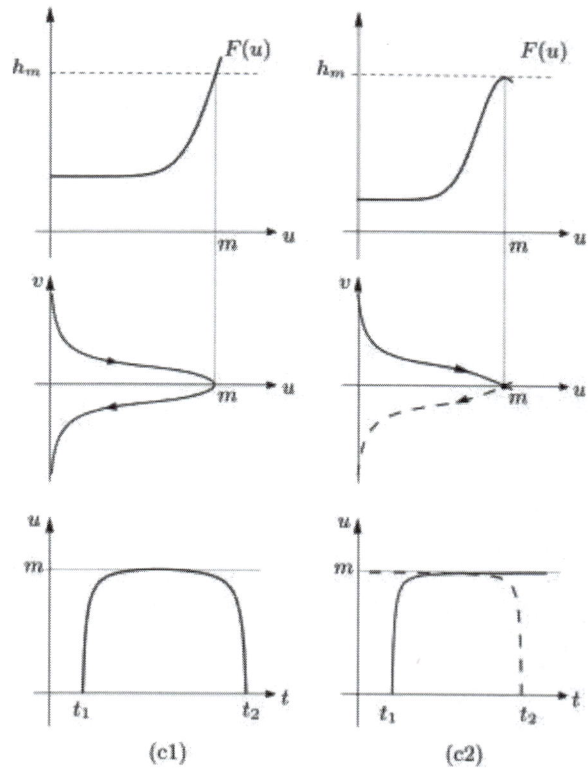

Figure 4. Strong solutions of (1) defined in a subset of R associated to curves γ_h with unbounded component v near the singular line $\{u=0\} \times R$. They provide elementary forms to construct TWS with cusps.

4. SINGULAR TRAVELING WAVE SOLUTIONS

This section is devoted to the characterization of singular TWS of Eq. (1). We are interested in singular TWS such that either $v_h^{\pm}(0)$ is well-defined for $h = h_0$, or $\lim_{u\to 0^+} v_h^{\pm}(u) = \pm\infty$ for $h \neq h_0$.

4.1. Peaked waves

A TWS of (1) given by a function $u : R \to R$ is called *peaked* if it is smooth except at a finite or countable number of points (*peaks*) $S = \{t_k \in \mathbb{R},\ k \in \mathbb{Z}\}$ where

$$0 \neq \lim_{t\to t_k^+} u'(t) = -\lim_{t\to t_k^-} u'(t) \neq \pm\infty.$$

The main result of this section is the following characterization:

Proposition 6.

Eq. (1) has peaked TWS if and only if

• $F'(0) < 0$ *and*

• *there exists* $m > 0$ *such that* $F(m) = F(0)$ *and* $F(u) < F(0)$ *for* $u \in (0, m)$.

These solutions are either

(i) *peaked periodic, with period* $T = \frac{2}{\sqrt{2}} \int_0^m \sqrt{\frac{u}{F(0) - F(u)}}\, du$, *if and only if in addition* $F'(m) \neq 0$, *or*

(ii) *peaked solitary if and only if in addition* $F'(m) = 0$.

These solutions are weak singular TWS and they are analytic except for a discontinuity in the first derivative at the peaks. Furthermore, the peaked solitary waves are symmetric with respect to their unique maximum and decay exponentially to zero at infinity. Peaked periodic solutions have a unique maximum and minimum per period and are symmetric with respect to these local extrema.

Proof.

Peaked TWS are compositions of the elementary forms studied in the case (a) of Section 3.2. They appear as a consequence of the existence of values $t \in R$ such that the solutions of (1) reach $u=0$ with non-vanishing one-sided derivative. These solutions are associated to the integral curves of (5) with energy $h=h_0$. The branches $(u, v_{h_0}^{\pm}(u))$ must be defined and bounded, intersecting the singular line $\{u=0\}$ at finite points $(0,+a)$ and $(0,-a)$ different from $(0,0)$, so that the corresponding solutions reach 0 with non-vanishing derivative. Therefore, there must exist $m>0$ such that $F(u)<F(0)$ for $u \in (0,m)$ and $F(m)=F(0)$, and additionally $F'(0)<0$. These two necessary conditions are also sufficient for the existence of peaks, as we will see in the construction of solutions in the proofs of (i) and (ii).

Suppose that $F'(m) \neq 0$, so $(m,0)$ is a regular point and therefore, by the symmetry of system (5) and by Eq. (10), there is an orbit of system (5) connecting $(0,+a)$ with $(0,-a)$ in finite time. This means that there exist $t_1, t_2 < \infty$ such that the first component of the solution of system (5), $u(t)$, is a strong solution of Eq. (1) (and Eq. (3) with $h=h_0$) with maximal interval of definition given by $t \in (t_1, t_2)$. This solution is the elementary form studied in the case (a1) in Section 3.2, hence $\lim_{t \to t_1^+} u(t) = 0$ and $\lim_{t \to t_1^+} \dot{u}(t) = a$. By symmetry, $\lim_{t \to t_2^-} u(t) = 0$, $\lim_{t \to t_2^-} \dot{u}(t) = -a$ and $u\left(\frac{t_1+t_2}{2}\right) = m$. Lemma 3 allows us to extend the above solution continuously to R by gluing together copies of it defined in the intervals (t_k, t_{k+1}) with $k \in Z$, where $t_k = t_1 + k(t_2 - t_1)$. This leads to the function

$$\tilde{u}(t) = \begin{cases} u\,(t - (k-1)(t_2 - t_1)) & \text{for } t \in (t_k, t_{k+1}), \\ 0 & \text{for } t = t_k, \end{cases} \tag{12}$$

which is periodic with period

$$T = t_2 - t_1 = \frac{2}{\sqrt{2}} \int_0^m \sqrt{\frac{u}{F(0) - F(u)}}\, du.$$

By Lemma 5(ii) the above integral is well defined, since $F'(m) \neq 0$. Notice also that $\lim_{t \to t_k} \tilde{u}(t) = 0$ and $\lim_{t \to t_k^{\pm}} \dot{\tilde{u}}(t) = \pm a$, so $\tilde{u}(t)$ is peaked periodic. Furthermore, the solution attains a unique maximum m at $t = (t_k + t_{k+1})/2$ and is symmetric with respect to it on each period. Observe that $\tilde{u} \in H^1_{loc}(\mathbb{R})$. Indeed, the function $\tilde{u}$ is bounded and analytic in $\mathbb{R} \setminus S$ where $S = \{t_k,\ k \in \mathbb{Z}\}$, and since $\lim_{t \to t_k} (\dot{\tilde{u}}(t))^2 = a^2$, also $|\dot{\tilde{u}}|^2$ is bounded in $\mathbb{R} \setminus S$. Recalling in addition that $\tilde{u}(t)$ is a strong solution of (1) on each interval (t_k, t_{k+1}), cf. (a1) in Section 3.2, we get that it is a TWS in the sense of Definition 1. By construction $\tilde{u}(t)$ satisfies Eq. (3) with $h = h_0$ except at the points in S where the equation holds in the limit (4). So according to Definition 2, it is a weak singular TWS of(1). Finally, observe that the solution $\tilde{u}$ constructed above is the only possible continuous continuation of $u(t)$ on R.

Suppose now that $F'(m) = 0$, so $(m, 0)$ is a critical point of (5). In this case there exist two orbits connecting $(m, 0)$ with $(0, +a)$ and $(0, -a)$, respectively, cf. the elementary forms of (a2) in Section 3.2. Again, Lemma 5(i) guarantees that any point on these orbits reaches $(0, \pm a)$ in finite time, but it takes an infinite time to reach $(m, 0)$. So there exist strong solutions $u_{\pm}(t)$ of (1) defined on (t_1, ∞) and $(-\infty, t_2)$, respectively. The non-uniqueness of solutions then allows us to choose $t_2 = t_1$ to construct the function

$$u(t) = \begin{cases} u_-(t) & \text{for } t \in (-\infty, t_1], \\ u_+(t) & \text{for } t \in (t_1, +\infty), \end{cases}$$

which is a peaked solitary wave defined on R. The same arguments as before show that u is in $H^1_{loc}(\mathbb{R})$, it is a weak solution of (1), and in particular a weak singular TWS of(1). Again, u is the only possible continuous continuation of u_- and u_+ onto R.

4.2. Solitary waves with compact support and associated composite waves

A TWS of (1) given by a function $u: R \to R$ has *compact support* if there exists $-\infty < t_1 < t_2 < \infty$ such that u is constant on $R \setminus [t_1, t_2]$ and u is non constant on $[t_1, t_2]$. We call $[t_1, t_2]$ the support of u. A solitary wave with compact support, or simply *compact solitary wave*, is a continuous TWS which has compact support and a unique extremum. We say that a TWS is a composite wave associated to a compact solitary wave $u(t)$ with support $[t_1, t_2]$, or simply a *composite wave of* $u(t)$, if it is obtained by gluing together copies of one compact solitary wave in such a way that the supports of each copy do not overlap. More precisely, $\tilde{u}(t)$ is a composite wave of a solitary wave $u(t)$ if there exists a collection of intervals $I_k = (t_1 - a_k(t_2 - t_1), t_2 - a_k(t_2 - t_1))$ for $a_k \in R$ and $k \in K$, where K is either Z or a finite collection of indices, with $I_k \cap I_j \neq \emptyset$ for all $k \neq j$, such that

$$\tilde{u}(t) = \begin{cases} u(t + a_k(t_2 - t_1)) & \text{if } t \in I_k \text{ for some } k \in K, \\ 0 & \text{if } t \in \mathbb{R} \setminus \bigcup_{k \in K} I_k. \end{cases}$$

The following result characterizes the compact solitary waves and composite waves.

Proposition 7.

Eq. (1) has compact solitary TWS and their associated composite solutions if and only if

- $F'(0) = 0$ *and* $F''(0) < 0$, *and*
- *there exists* $m > 0$ *such that* $F(m) = F(0)$, $F(u) < 0$ *for* $u \in (0, m)$ *and* $F'(m) \neq 0$.

These solutions are either compact solitary waves, composite multi-bump solutions with compact support or composite waves with non-compact support that can be either aperiodic or periodic with arbitrary period. These solutions are strong singular TWS, they are $C^1(R)$ *and piecewise analytic. Furthermore, the compact solitary waves are symmetric with respect to their unique maximum.*

Proof.

Compact solitary TWS appear when there exist values $t \in R$ such that the solutions of (1)reach $u=0$ with vanishing one-sided derivative. These solutions are associated to the integral curves of (5) with energy $h=h_0$, and as before, the branches $(u, v_{h_0}^{\pm}(u))$ must be defined and bounded such that they intersect the singular line $\{u=0\}\times R$ at $(0,0)$ in finite time. This last fact guarantees that the corresponding solution of (1) reaches $u=0$ with vanishing derivative in finite time.

As studied in the case (b) of Section 3.2, the existence and boundedness of such curves is guaranteed if there exists $m>0$ such that $F(u)<F(0)$ for $u\in(0,m)$ and $F(m)=F(0)$. The intersection of the curves $(u, v_{h_0}^{\pm}(u))$ with the singular line $\{u=0\}\times R$ at $(0,0)$ is ensured if $F'(0)=0$, since in this case

$$\lim_{u\to 0+} v_{h_0}^{\pm}(u) = \lim_{u\to 0+} \pm\sqrt{\frac{-2\,F''(u)u^2+o(u^3)}{u}} = 0. \quad (13)$$

By Lemma 5(i) and (ii), the orbits of (5) connect $(0,0)$ with $(m,0)$ in finite time if $F''(0)<0$ and $F'(m)>0$. Under these conditions there exist $t_1,t_2\in R$ such that the first component of the solution of system (5), $u(t)$, is a strong solution of (1) (and also (3)with $h=h_0$) in the maximal interval of definition $t\in(t_1,t_2)$. Moreover, $\lim_{t\to t_1^+} u(t) = 0$ and by symmetry $\lim_{t\to t_2^-} u(t) = 0$ and $u\left(\frac{t_1+t_2}{2}\right) = m$ is the unique maximum. This solution $u(t)$ is the elementary form that appears in the case (b1) of Section 3.2.

In particular, the above conditions imply that Eq. (1) is not unique at $u=0$, and that $u(t)\equiv 0$ is also a solution. Therefore it is possible to obtain a continuous continuation of $u(t)$ with compact support in R. We define the function

$$\tilde{u}(t) = \begin{cases} u(t) & \text{for } t \in (t_1, t_2), \\ 0 & \text{for } t \in \mathbb{R}\setminus(t_1, t_2). \end{cases}$$

Observe that $\tilde{u}(t) \in H^1_{loc}(\mathbb{R})$ and it is analytic in $\mathbb{R}\setminus\{t_1, t_2\}$. Furthermore, $\lim_{t \to t_1^+} \dot{u}(t) = 0$ and $\lim_{t \to t_2^-} \dot{u}(t) = 0$ in view of (13). Hence $\tilde{u}(t) \in C^1(\mathbb{R})$, and by construction it is a strong singular TWS of (1). In a similar way as above, we can also construct the following composite waves: multi-bump waves with compact support by gluing together a finite number of copies of $u(t)$, and waves with non-compact support that can be either aperiodic or periodic with arbitrary period. All these solutions are strong singular TWS of (1), they are piecewise analytic and at most $C^1(R)$, cf. Remark 9.

The sufficiency part of the proof follows from the fact that if F satisfies the stated conditions, then it is possible to find the desired solutions following the above construction.

As a direct consequence of Proposition 6 and Proposition 7, we obtain that it is not possible to find a peaked wave with compact support.

Corollary 8.

Eq. (1) *cannot admit peaked and compactly supported TWS for the same* F. *In particular, a peaked TWS cannot have compact support, and conversely, a TWS with compact support cannot have peaks.*

Remark 9.

In view of Lemma 5, the conditions of Proposition 7 guarantee the existence of a homoclinic orbit of system (5) where every point on the orbit is connected to $(0,0)$ in a finite time. They also account for the fact that the compactly supported solutions are at most C^1. Indeed, notice that under the conditions of Proposition 7 we have that

$$\dot{u}(t) = v_{h_0}^{\pm} = \pm\sqrt{-2F''(u)u + o(u^2)}$$

and

$$\begin{aligned}\ddot{u}(t) &= \pm\frac{-2F^{(3)}(u)u\dot{u} - 2F''(u)\dot{u} + o(u)\dot{u}}{2(\pm\dot{u})}\\ &= -F^{(3)}(u)u - F''(u) + o(u).\end{aligned}$$

Therefore,

$$\lim_{t\to t_1^-}\ddot{u}(t) = -F''(u) \neq 0 = \lim_{t\to t_1^+}\ddot{u}(t)$$

$$\lim_{t\to t_2^+}\ddot{u}(t) = -F''(u) \neq 0 = \lim_{t\to t_1^-}\ddot{u}(t), \tag{14}$$

and hence the continuation to $u(t)\equiv 0$ is not C^2 since $F''(u)<0$. If we were to demand a C^2-continuation, this would necessarily require the second derivative of F to vanish. As a consequence, the solutions would loose the compactness property, since in view of Lemma 5 the existence time of the loop would become infinite.

Remark 10.

Observe that a different kind of composition of locally defined solutions may also be obtained under conditions stated in Proposition 7 but with $F'(0)<0$. In this case

$$\lim_{u\to 0+} v_{h_0}^{\pm}(u) = \lim_{u\to 0+} \pm\sqrt{\frac{-2\,F'(u)u + o(u^2)}{u}} = \pm\sqrt{-2F'(0)} =: \pm a \neq 0,$$

and hence $\lim_{t\to t_1^+}\dot{u}(t) = a$ and $\lim_{t\to t_2^-}\dot{u}(t) = -a$.
However, $u(t)\equiv 0$ is not a solution of (1) under this assumption on F and therefore the only possible continuous continuation is given by the periodic function (12). But this is not a composition of a compact solitary wave, i.e. not a composite wave as defined at the beginning of this section.

4.3. Fronts with finite-time decay and plateau-shaped waves

Proposition 11.

Eq. (1) *has fronts, solitary and plateau-shaped singular TWS solutions if and only if*

- $F'(0)=0$ *and* $F''(0)<0$, *and*
- *there exists* $m>0$ *such that* $F(m)=F(0)$, $F(u)<0$ *for* $u\in(0,m)$ *and* $F'(m)=0$.

The solitary waves are strong solutions of (1) *and therefore analytic. The fronts and plateau-shaped waves are strong singular TWS which are at most* C^1 *and piecewise analytic. Moreover, the fronts have finite-time decay on one side.*

Proof.

Under the present hypothesis, there exist orbits of system (5) corresponding to the branches $(u, v_{h_0}^{\pm}(u))$, which connect $(0,0)$ with the critical point $(m,0)$. By Lemma 5, any point on these orbits is connected to $(0,0)$ in finite time and to $(m,0)$ in infinite time. They yield two strong solutions $u_{\pm}(t)$ with maximal interval of definition (t_1,∞) and $(-\infty,t_2)$, respectively, which are given by the elementary forms of the case (b2) of Section 3.2. By gluing them together with the solution $u(t)\equiv 0$ we get the following solutions: fronts given by

$$\tilde{u}_-(t) = \begin{cases} u_-(t) & \text{for } t \in (-\infty, t_1), \\ 0 & \text{for } t \in [t_1, \infty), \end{cases} \quad \text{and} \quad \tilde{u}_+(t) = \begin{cases} 0 & \text{for } t \in (-\infty, t_2], \\ u_+(t) & \text{for } t \in (t_2, \infty). \end{cases}$$

Setting $t_1=t_2$ we get a solitary wave given by

$$u(t) = \begin{cases} u_-(t) & \text{for } t \in (-\infty, t_1), \\ 0 & \text{for } t = t_1, \\ u_+(t) & \text{for } t \in (t_1, \infty), \end{cases}$$

which is symmetric with respect to its unique maximum. Choosing $t_1 < t_2$ we get a plateau-shaped solitary wave given by

$$u(t) = \begin{cases} u_-(t) & \text{for } t \in (-\infty, t_2), \\ 0 & \text{for } t \in [t_2, t_1], \\ u_+(t) & \text{for } t \in (t_1, \infty). \end{cases}$$

The front and plateau-shaped solutions are analytic in $R \setminus \{t_i\}$ and satisfy Eq. (3) for $h = h_0$, hence they are strong singular TWS of (1). Furthermore, they are at most $C^1(R)$ for the same reason explained in Remark 9. The solitary wave solutions, however, are in fact strong solutions of (1) on R since $\ddot{u}(t)$ is defined everywhere, cf. (14), and therefore they are analytic on R.

We emphasize that the fronts described above are not the classical smooth fronts with exponential decay on both ends, but they decay in finite time on one end.

4.4. Cusped waves

A TWS of (1) given by a function $u: R \to R$ is called *cusped* if it is smooth except at a finite or countable number of points (*cusps*) $S = \{t_k \in \mathbb{R},\ k \in \mathbb{Z}\}$ where

$$\lim_{t \to t_k^+} u'(t) = - \lim_{t \to t_k^-} u'(t) = \pm\infty.$$

Proposition 12.

Eq. (1) *has cusped TWS if and only if there exists* $m > 0$ *such that* $F(m) - F(u) > 0$ *for all* $u \in [0, m)$. *These solutions are either*

(i) *cusped periodic with period* $T = \frac{2}{\sqrt{2}} \int_0^m \sqrt{\frac{u}{F(m) - F(u)}} du$ *if and only if in addition* $F'(m) \neq 0$, *or*

(ii) *cusped solitary if and only if in addition* $F'(m) = 0$.

These solutions are weak singular TWS and they are analytic except for a discontinuity in the first derivative at the peaks. Furthermore, the cusped solitary waves are symmetric with respect to their unique maximum and decay exponentially to zero at infinity. Cusped periodic solutions have a unique maximum and minimum per period and are symmetric with respect to these local extrema.

Proof.

Cusped TWS correspond to orbits of (5) which are defined by the curves $(u, v_h^{\pm}(u))$ with $h \neq h_0$ satisfying

$$\lim_{u \to 0+} v_h^{\pm}(u) = \lim_{u \to 0+} \pm\sqrt{2\frac{h - F(u)}{u}} = \pm\infty.$$

Therefore, a necessary condition for the appearance of cusps is that there exists $m > 0$ such that $F(m) - F(u) > 0$ for $u \in [0, m)$. In particular, this implies that $h_m := F(m) > F(0)$. As shown in the case (c) of Section 3.2 the orbits of (5) passing through the points $(\bar{u}, v_{h_m}^{\pm}(\bar{u}))$ are unbounded in the component v but approach the singular line $u = 0$ at infinity in a finite time. We will see in the construction of solutions in the proofs of (i) and (ii) below that the necessary condition deduced above is also sufficient for the existence of cusped TWS.

(i) If $F'(m) \neq 0$ then the point $(m, 0)$ connecting the two branches $(u, v_{h_m}^{\pm}(u))$ is regular, so by Eq. (11) there exist $t_1, t_2 < \infty$ such that the first component of the solution of system(5) $u(t)$, is a strong solution of (1) (and also of (3) with $h = h_m$), with maximal interval of definition $t \in (t_1, t_2)$, which is given by the elementary form that appears in case (c1) of Section 3.2. Notice that $u(t) \equiv 0$ is not a solution of (3) for $h = h_m > 0$. Therefore, the only possible continuous continuation of $u(t)$ preserving the energy is given by the periodic function

$$\tilde{u}(t) = \begin{cases} u\left(t - (k-1)(t_2 - t_1)\right) & \text{for } t \in (t_k, t_{k+1}), \\ 0 & \text{for } t = t_k, \end{cases}$$

where $t_k = t_1 + k(t_2 - t_1)$ for $k \in Z$, which is periodic with period

$$T = t_2 - t_1 = \frac{2}{\sqrt{2}} \int_0^m \sqrt{\frac{u}{F(m) - F(u)}} du.$$

Observe now that integrating system (5), and denoting $u_\varepsilon := u(t_1 + \varepsilon)$ we get,

$$\int_{t_1}^{t_1+\varepsilon} \dot{u}^2 dt = \int_0^{u_\varepsilon} v^2 \frac{du}{v} = \int_0^{u_\varepsilon} \sqrt{\frac{2(F(m) - F(u))}{u}} du.$$

Since $F(m) - F(u) > 0$ for all $u \in [0, u_\varepsilon]$ and recalling the criterion used in Lemma 5, we find that $\lim_{u \to 0^+} \sqrt{(F(m) - F(u))/u} \cdot \sqrt{u} = A$ with $A \neq 0$ and $A \neq \infty$. Therefore,

$$\int_0^{u_\varepsilon} \sqrt{\frac{2(F(m) - F(u))}{u}} du < \infty,$$

and hence $\int_{t_1}^{t_1+\varepsilon} \dot{u}^2 dt < \infty$. Analogously $\int_{t_2-\varepsilon}^{t_2} \dot{u}^2 dt < \infty$ and hence $\int_{t_k-\varepsilon}^{t_k+\varepsilon} (\dot{\tilde{u}})^2 dt$ is convergent for every t_k so that for each compact $K \subset R$ we have

$$\int_K (\dot{\tilde{u}})^2 dt < \infty.$$

Since $\tilde{u}(t)$ is bounded as well, this solution is in $H^1_{loc}(\mathbb{R})$, and analytic on $R \setminus S$ where $S = \{t_k \in \mathbb{R},\ k \in \mathbb{Z}\}$. Finally observe that $\tilde{u}(t)$ is a weak singular TWS of (1), because

$$\lim_{t \to t_k} \frac{u(\dot{u})^2}{2} + F(u) = u\, \frac{F(m) - F(u)}{u} + F(u) = F(m) = h_m.$$

(ii) If $F'(m)=0$ then the orbits corresponding to $(u, v_{h_m}^{\pm})$ approach the singular line $\{u=0\}$ in finite time, but it takes an infinite time to reach $(m,0)$. So there exist two strong solutions of (1) $u_{\pm}(t)$ given by the elementary forms of the case (c2) of Section 3.2, defined on the maximal intervals $(t_1,+\infty)$ and $(-\infty,t_2)$, respectively. The only way to construct a continuous continuation in R preserving the energy is by choosing $t_2=t_1$ and gluing together the corresponding solutions $u_{\pm}$. These considerations lead us to define the function

$$\tilde{u}(t) = \begin{cases} u_-(t) & \text{for } t \in (-\infty, t_1), \\ 0 & \text{for } t = t_1 \\ u_+(t) & \text{for } t \in (t_1, +\infty), \end{cases}$$

which is a cusped solitary TWS defined in R. The same arguments as in the proof of statement (a) show that $\tilde{u}$ is a weak singular TWS of (1).

4.5. Exhaustivity of the characterization

Observe that the elementary forms presented in Section 3.2 capture all the strong solutions of Eq. (1) reaching or tending to $\{u=0\}$, whose maximal interval of definition is not R. In addition, the singular solutions described in the preceding Section 4 cover all possible continuous extensions to R on the same energy level, using these elementary forms and the constant function $u(t)\equiv 0$ whenever it is a solution. In consequence, our characterization of singular TWS for Eq. (1) given
in Proposition 6, Proposition 7,Proposition 11 and Proposition 12 is exhaustive.

5. APPLICATION TO SHALLOW WATER EQUATIONS

The aim of this section is to demonstrate the applicability of the propositions developed in the preceding sections. We exemplify our approach by studying the equation for surface waves of moderate amplitude in shallow water and the Camassa–Holm equation. In particular, we show how the different types of singular TWS can be obtained varying the energy of the corresponding Hamiltonian systems.

This approach may be applied to study singular TWS of a variety of other equations, for example a class of nonlinear wave equations related to the inviscid Burgers' equation and Camassa–Holm equation studied in [12], the family of equations analyzed in [13], and a generalization of the Camassa–Holm equation studied in [14]. In the latter paper, the authors conclude with a conjecture on the non-existence of peaked solitary solutions when a certain parameter becomes non-positive. In view of the results in Section 4 we are able to give an affirmative answer.

5.1. Surface waves of moderate amplitude in shallow water

In this section we study singular TWS of the equation for surface waves of moderate amplitude in shallow water,

$$u_t + u_x + 6uu_x - 6u^2u_x + 12u^3u_x + u_{xxx} - u_{xxt} + 14uu_{xxx} + 28u_xu_{xx} = 0, \tag{15}$$

which was first derived by Johnson [15], whose considerations were extended by Constantin and Lannes [4]. We refer to [16] for a first study of smooth solitary waves and to [8] for a more extensive characterization of TWS of Eq. (15). We introduce the traveling wave Ansatz $u(x,t) = u(x - ct)$ and integrate once to obtain

$$u''\left(u + \frac{1+c}{14}\right) + \frac{1}{2}(u')^2 + K + (1-c)u + 3u^2 - 2u^3 + 3u^4 = 0, \tag{16}$$

for some constant $K \in R$. Notice that after the change of variables $u \mapsto u - \frac{1+c}{14}$ the above equation is of the form (1), with F a suitable polynomial in u depending on the parameters c and K. The relation between the parameters and the qualitative properties of F is studied in detail in [8]. In particular, it is observed that F has either no extremum or there are two extrema which we denote by p_1 (the local maximum) and p_2 (the local minimum of F) such that $p_1 < p_2$. Let $h_i = F(p_i)$ for $i = 1, 2$. We distinguish different cases depending on the position of p_1 and p_2 with respect to $u = 0$, and the sign of $h_0 - h_2$. Taking into account these cases and the characterization given in Proposition 6, Proposition 7 and Proposition 12 we obtain the following types of bounded singular TWS for Eq. (15) varying the energy. We point out that, as a consequence of Corollary 8, compact solitary waves and peaked waves cannot coexist within a single case. To give an example, Fig. 5 shows the different TWS of Eq. (15) that appear for different energy levels when $0<p_1<p_2$ and $h_2>h_0$.

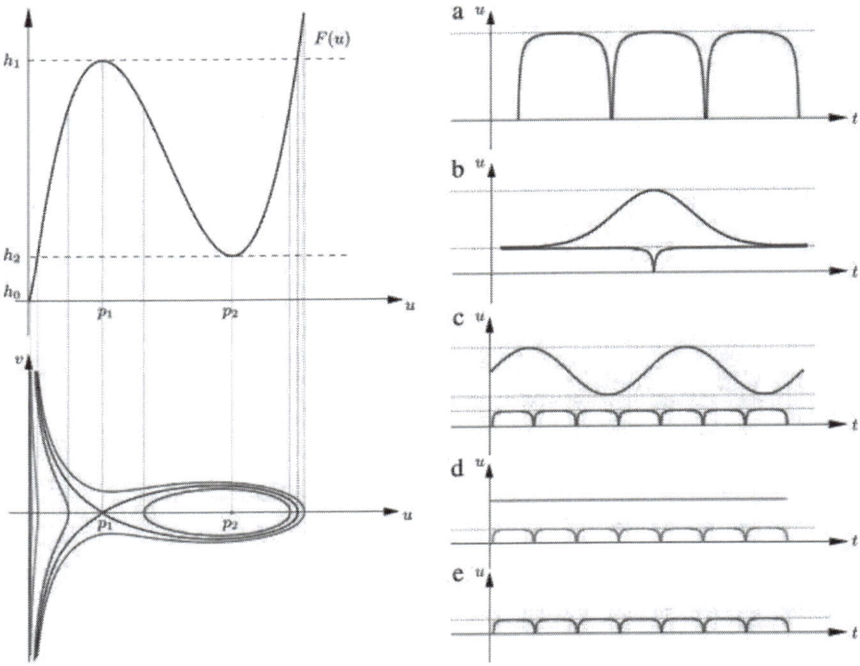

Figure 5. Singular TWS of Eq. (15) varying the energy level h in the case $0<p_1<p_2$ and $h_2>h_0$. (a) cusped periodic waves for $h>h_1$; (b) cusped and smooth solitary waves for $h=h_1$; (c) cusped and smooth periodic waves for $h_1>h>h_2$; (d) cusped periodic waves and constant solutions for $h=h_2$; (e) cusped periodic waves for $h_2>h>h_0$.

Energy/case	$0<p_1<p_2$ and $h_2>h_0$	Energy/case	$0<p_1<p_2$ and $h_2<h_0$
$h>h_1$	Cusped periodic	$h>h_1$	Cusped periodic
$h=h_1$	Cusped and smooth solitary	$h=h_1$	Cusped and smooth solitary
$h_1>h>h_2$	Cusped and smooth periodic	$h_1>h>h_0$	Cusped and smooth periodic
$h=h_2$	Cusped periodic and constant	$h_0\geq h>h_2$	Smooth periodic
$h_2>h>h_0$	Cusped periodic	$h=h_2$	Constant
$h\leq h_0$		$h\leq h_2$	

Energy/case	$0 < p_1 < p_2$ and $h_2 = h_0$	Energy/case	$0 = p_1 < p_2$
$h > h_1$	Cusped periodic	$h > h_1$	Cusped periodic
$h = h_1$	Cusped and smooth solitary	$h = h_1 = h_0$	Compact solitary and composite
$h_1 > h > h_0 = h_2$	Cusped and smooth periodic	$h_1 > h > h_2$	Smooth periodic
$h = h_0$	Constant	$h = h_2$	Constant
$h < h_0$		$h < h_2$	

Energy/case	$p_1 < 0 < p_2$	$p_1 < p_2 = 0$	$p_1 < p_2 < 0$
$h > h_0$	Cusped periodic	Cusped periodic	Cusped periodic
$h = h_0$	Peaked periodic	Constant	
$h_0 > h > h_2$	Smooth periodic		
$h = h_2$	Constant		
$h < h_2$			

5.2. The Camassa-Holm equation

In the present section we will study singular TWS of the Camassa-Holm equation (CH)

$$u_t + 2\kappa\, u_x - u_{txx} + 3\,u\,u_x = 2\,u_x u_{xx} + u\,u_{xxx}, \quad (17)$$

for $x \in R$, $t > 0$ and $\kappa \in R$, which was introduced in the context of water waves by Camassa and Holm [1]. For a classification of weak traveling wave solutions of the Camassa-Holm equation we refer to [17]. Proceeding as in the previous section we introduce the traveling wave Ansatz $u(x,t) = u(x - ct)$. Integrating once Eq. (17) takes the form

$$u''(u-c) + \frac{(u')^2}{2} + r + (c - 2\kappa)\,u - \frac{3}{2}u^2 = 0,$$

where r is a constant of integration. The change of variables

$$w = u - c,$$

transforms the above equation to the form (1) with

$$F(w) = A\,w + B\,w^2 - \frac{1}{2}w^3, \tag{18}$$

where $A = r - 2\kappa c - \frac{1}{2}c^2$ and $B=-(c+\kappa)$. $F(w)$ is a third order polynomial which satisfies $F(0)=0$, it has at most three roots

$$w = 0 \quad \text{and} \quad w = B \pm \sqrt{B^2 + 2A},$$

and at most two extrema

$$p_i = \frac{2B + (-1)^i\sqrt{4B^2 + 6A}}{3}, \quad i = 1, 2.$$

We may assume that $B \geq 0$ since otherwise the change of variables $(\hat{w}, \hat{v}) = -(w, v)$yields this situation (this is equivalent to considering system (5) only for $u \geq 0$). Note that F does not have any extremum when $4B^2+6A \leq 0$, and it has two distinct extrema otherwise. In the latter case, p_1 is the local minimum and $p_2 > 0$ the local maximum (with $F''(p_2)<0$). We denote $h_1=F(p_1)$ and $h_2=F(p_2)$, and distinguish between the following cases:

Case					
(i)	$A > 0$			$p_1 < 0 < p_2$	$h_1 < 0 < h_2$
(ii)	$A = 0,$	$B > 0$		$p_1 = 0 < p_2$	$h_1 = 0 < h_2$
(iii)	$A < 0,$	$4B^2 + 6A > 0,$	$B^2 + 2A > 0$	$0 < p_1 < p_2$	$h_1 < 0 < h_2$
(iv)	$A < 0,$	$4B^2 + 6A > 0,$	$B^2 + 2A = 0$	$0 < p_1 < p_2$	$h_1 < h_2 = 0$
(v)	$A < 0,$	$4B^2 + 6A > 0,$	$B^2 + 2A < 0$	$0 < p_1 < p_2$	$h_1 < h_2 < 0$
(vi)	$A < 0,$	$4B^2 + 6A \leq 0$			

Taking into account the cases described above and the classification given inProposition 6 and Proposition 12 we obtain the following types of singular TWS varying the energy level h:

Energy/case	(i)	(ii)	(iii)	(iv)
$h < h_1$				
$h = h_1$			Constant	Constant
$h_1 < h < 0$			Smooth periodic	Smooth periodic
$h = 0$		Constant	Peaked periodic	Peaked solitary
$0 < h < h_2$	Cusped periodic	Cusped periodic	Cusped periodic	
$h = h_2$	Cusped solitary	Cusped solitary	Cusped solitary	
$h > h_2$				

Energy/case	(v)
$h < h_1$	
$h = h_1$	Constant
$h_1 < h < h_2 < 0$	Smooth periodic
$h = h_2$	Smooth solitary
$h > h_2$	

5.3. Generalized Camassa-Holm equation

In [14] the authors study peaked solitary and periodic cusped traveling wave solutions of a generalization of the CH equation of the form

equation(19)

$$u_t + 2\kappa u_x - u_{txx} + a u u_x = 2 u_x u_{xx} + u u_{xxx},$$

where $a \in R$ is an additional parameter. At the end of their paper they state the following conjecture: "If the parameter $a \leq 0$, then Eq. (19) has no peaked solitary wave solution". Using the approach developed in the preceding sections it is easy to see that this assertion is true. Proceeding as with the CH above we introduce the traveling wave Ansatz and integrate once to find that after the change of variables $w = u - c$ we obtain

$$w'' w + \frac{(w')^2}{2} + F'(w) = 0,$$

which is an equation of the form (1) with

$$F(w) = A\,w + B\,w^2 - \frac{a}{6}w^3,$$

where $A = r + (1 - \frac{a}{2})c^2 - 2\kappa c$ and $B = (1 - a)\frac{c}{2} - \kappa$. Our analysis (cf. Proposition 6) shows that an equation of this form has peaked solitary TWS if and only if $F'(0) < 0$ and there exists $m > 0$ such that $F(m) = F(0)$ with $F(u) < F(0)$ for $u \in (0, m)$ and $F'(m) = 0$. Assuming that $a < 0$, we see that $F(w) \to \pm\infty$ as $w \to \pm\infty$. This contradicts the conditions for the existence of peaked solitary solutions stated above. Indeed, if Eq. (19) had peaked solitary TWS, then $F'(0) = A < 0$. Thus, F would have a maximum to the left and a minimum to the right of $w = 0$. Hence there exists $m > 0$ such that $F(m) = F(0)$, but $F'(m) \neq 0$ since F has at most two extrema. For $a = 0$ the situation is similar. This shows that the conjecture is true.

ACKNOWLEDGMENTS

The first author is supported by the project "Dynamical Systems Methods in Hydrodynamics", project number J3452, of the Austrian Science Fund (FWF). The second author is partially supported by Ministry of Economy and Competitiveness of the Spanish Government through grant DPI2011-25822 and grant 2014-SGR-859 from AGAUR, Generalitat de Catalunya.

REFERENCES

1. R. Camassa, D.D. Holm, An integrable shallow water equation with peaked solitons, Phys. Rev. Lett. 71 (11) (1993) 1661–1664.

2. R. Camassa, D.D. Holm, J.M. Hyman, A new integrable shallow water equation, Adv. Appl. Mech. 31 (31) (1994) 1–33.

3. B. Fuchssteiner, A.S. Fokas, Symplectic structures, their B¨acklund transformations and hereditary symmetries, Physica D4 (1981) 821–831.

4. A. Constantin, D. Lannes, The hydrodynamical relevance of the Camassa–Holm and Degasperis–Procesi equations, Arch. Ration. Mech. Anal. 192 (2009) 165–186.

5. N. Duruk Mutluba, s, On the Cauchy problem for a model equation for shallow water waves of moderate amplitude, Nonlinear Anal. RWA 14 (5) (2013) 2022–2026.

6. N. Duruk Mutluba, s, Local well-posedness and wave breaking results for periodic solutions of a shallow water equation for waves of moderate amplitude, Nonlinear Anal. Theory Methods Appl. 97 (2014) 145–154.

7. N. Duruk Mutluba, s, A. Geyer, B.V. Matioc, Non-uniform continuity of the flow map for an evolution equation modeling shallow water waves of moderate amplitude, Nonlinear Anal. RWA 17 (2014) 322–331.

8. A. Gasull, A. Geyer, Traveling surface waves of moderate amplitude in shallow water, Nonlinear Anal. Theory Methods Appl. 102 (2014) 105–119.

9. M. Ehrnstr¨om, H. Holden, X. Raynaud, Symmetric waves are traveling waves, Int. Math. Res. Not. 2009 (24) (2009) 4578–4596.

10. J. Lenells, Traveling wave solutions of the Camassa–Holm equation, J. Differential Equations 217 (2) (2005) 393–430.

11. V. Arnol'd, Ordinary Differential Equations, Springer, Berlin, 1992.

12. J. Lenells, Classification of traveling waves for a class of nonlinear wave equations, J. Dynam. Differential Equations 18 (2) (2006) 381–391.

13. Y.A. Li, P.J. Olver, Convergence of solitary-wave solutions in a perturbed bi-Hamiltonian dynamical system. I. Compactons and peakons, Discrete Contin. Dyn. Syst. 3 (3) (1997) 419–432.

14. Z. Liu, R. Wang, Z. Jing, Peaked wave solutions of Camassa–Holm equation, Chaos Solitons Fractals 19 (2004) 77–92.

15. [15] R.S. Johnson, Camassa–Holm, Korteweg–de Vries and related models for water waves, J. Fluid Mech. 455 (2002) 63–82.

16. A. Geyer, Solitary traveling water waves of moderate amplitude, J. Nonlinear Math. Phys. 19 (supp.01) (2012) 1240010. 12.

17. J. Lenells, Traveling wave solutions of the Camassa–Holm and Korteweg–de Vries equations, J. Nonlinear Math. Phys. 11 (4) (2004) 508–520.

2

CHAPTER

Solitary Wave Solutions Of The Generalized Two-Component Hunter–Saxton System

Byungsoo Moon

Department of Mathematics, University of Texas at Arlington, Arlington, TX 76019-0408, USA

ABSTRACT

The existence of solitary wave solutions of the generalized two-component Hunter–Saxton system is determined. It is also shown that there are peaked and cusped solitary waves with singularities among those smooth solitary wave solutions.

Keywords: Generalized two-component; Hunter–Saxton system; Solitary waves

1. INTRODUCTION

Consider the generalized two-component Hunter–Saxton system (gHS2)

$$\begin{cases} u_{txx} + 2\sigma u_x u_{xx} + \sigma u u_{xxx} - \rho\rho_x + A u_x = 0, \\ \rho_t + (\rho u)_x = 0, \end{cases} \tag{1.1}$$

where $\sigma \in R$ is the new free parameter, and $A \geq 0$. System (1.1) is the short-wave (or high-frequency) limit

$$(t, x) \mapsto (\epsilon t, \epsilon x), \quad \epsilon \to 0$$

of the generalized two-component Camassa–Holm system (gCH2) established in [1]which can be derived from shallow water theory with nonzero constant vorticity by using Ivanov's modeling approach [2],

$$\begin{cases} u_t - u_{txx} - A u_x + 3 u u_x - \sigma(2 u_x u_{xx} + u u_{xxx}) + \rho\rho_x = 0, \\ \rho_t + (\rho u)_x = 0, \end{cases}$$

where $u(t,x)$ represents the horizontal velocity of the fluid, and $\rho(t,x)$ is related to the free surface elevation from equilibrium (or scalar density) with the boundary assumption, $u \to 0$, $\rho \to 1$ as $|x| \to \infty$. The parameter $A > 0$ characterizes a linear underlying shear flow so that the two-component CH system models wave–current interactions [3] and [4]. The real dimensionless constant σ is a parameter which provides the competition, or balance, in fluid convection between nonlinear steepening and amplification due to stretching.

When $(\sigma, A) = (1, 0)$, (1.1) becomes the two-component Hunter–Saxton system. Mathematical properties of the system have been also studied further in many works. cf. [5], [6], [7], [8], [9], [10] and [11] and references therein. The two-component Hunter–Saxton system is formally integrable [12] with a bi-Hamiltonian structure—it can be written as a compatibility condition of two linear systems (Lax pair) with a spectral parameter ζ:

$$\Psi_{xx} = (-\zeta^2 \rho^2 + \zeta m)\Psi$$

$$\Psi_t = \left(\frac{1}{2\zeta} - u\right)\Psi_x + \frac{1}{2}u_x\Psi, \quad m = -u_{xx}.$$

If $(\sigma, A) = (1, 0)$ and $\rho \equiv 0$, then (1.1) reduces to the Hunter–Saxton equation (HS)[13],

$$u_{txx} + 2u_x u_{xx} + uu_{xxx} = 0,$$

which is modeling the propagation of weakly nonlinear orientation waves in a massive nematic liquid crystal (see Hunter and Saxton [13] for a derivation, and also [14], [15],[16], [17] and [18]). The Hunter–Saxton equation also arises in a different physical context as the high-frequency limit [19] and [20] of the Camassa–Holm equation for shallow water waves [21], [22] and [23] (see Ionescu-Kruse [24] for the alternative derivation) and a re-expression of the geodesic flow on the diffeomorphism group of the circle with a bi-Hamiltonian structure [13] and [25] which is completely integrable [14] and [20].

Our goal is to study solitary wave solutions of (1.1) having the form

$$(u(t,x), \rho(t,x)) = (\varphi(x - ct), \rho(x - ct)), \quad c \in \mathbb{R}$$

for some $\varphi, \rho : R \to R$ such that $\varphi \to 0,\ \rho \to 1$ as $|x| \to \infty$. Using a natural weak formulation of the generalized two-component Hunter–Saxton system, we will establish exactly in what sense the peaked and cusped solitary waves are solutions. It was shown in [26] that the μHS equation does not have solitary wave solutions on the line, but the periodic μHS equation has cusped and anticusped periodic traveling wave solutions. In [6], the authors considered the two-component HS equation for the particular choice of the parameter $(\sigma, A) = (1, 0)$. They showed that there exists a smooth periodic solution traveling wave of the two-component Hunter–Saxton equation and mentioned that Hunter–Saxton equation admits no bounded traveling waves. Moreover, when $\sigma = 1, \rho \equiv 0$, and $A \neq 0$ system (1.1) exhibits both smooth as well as cusped traveling waves [27] and the cusped soliton solutions of (1.1) in this case were investigated in the context of algebraic and complex geometry in [28]. However it is unclear whether the generalized two-component Hunter–Saxton system (1.1) has solitary waves with singularities. We show here peaked solitary waves exist when $\sigma>1$. It should be pointed out that the peaked traveling waves capture a feature of the water waves of great height (solutions of the largest amplitude to the governing equations for water waves) [29], [30] and [31]. Moreover, the shape of some peakons is stable under small perturbations, making these waves recognizable physically [32],[33] and [34]. We also provide an implicit formula for the peaked solitary waves. However whether these peaked solitary waves are solitons still remains open. It is worth mentioning that for the Camassa–Holm equation the smooth solitary waves are solitons [35].

A special case of system (1.1) is when $\sigma=0$. In [36], the authors showed that the solutions of (1.1) are global in time. We show that the same results hold in the system case.

The remainder of the paper is organized as follows. In Section 2, we derive a weak formulation of (1.1) convenient for solitary wave solutions. In Section 3 we determine the existence of solitary wave solutions of (1.1) for $\sigma=0$. In Section 4, we classify the solitary waves of (1.1) for $\sigma\neq 0$. In particular we show the existence of peaked solitary waves for $\sigma>1$.

2. PRELIMINARIES

Consider solitary wave solutions of (1.1) with $A>0$ having the form

$$(u(t,x),\rho(t,x))=(\varphi(x-ct),\rho(x-ct)),\quad c\in\mathbb{R}$$

for some $\varphi,\rho:R\to R$ such that $\varphi\to 0,\ \rho\to 1$ as $|x|\to\infty$. With the assumption $\rho\to 1$ as $|x|\to\infty$, we can define $\rho=1+\eta$ with $\eta\to 0$ as $|x|\to\infty$ and hence we can rewrite system (1.1) as

$$\begin{cases} u_{txx}+2\sigma u_x u_{xx}+\sigma u u_{xxx}-(1+\eta)\eta_x+Au_x=0,\\ \eta_t+((1+\eta)u)_x=0. \end{cases} \tag{2.1}$$

Now we give the definition of solitary waves of (2.1).

Definition 2.1.

A solitary wave of (2.1) is a nontrivial traveling wave solution of (2.1) of the form $(\varphi(x-ct),\eta(x-ct)) \in H^1(R) \times H^1(R)$ with $c \in R$ and φ, η vanishing at infinity along with the first and second derivatives of φ and η.

Integrating (2.1) over $(-\infty, x]$, and taking into account $\eta(x), \varphi(x), \varphi_x(x)$, and $\varphi_{xx}(x) \to 0$ as $|x| \to \infty$, one finds that $\varphi(x)$ and $\eta(x)$ must satisfy

$$\begin{cases} -c\varphi_{xx} + \sigma\varphi\varphi_{xx} + \frac{\sigma}{2}\varphi_x^2 - \frac{1}{2}(\eta+1)^2 + A\varphi + \frac{1}{2} = 0, \\ -c\eta + (1+\eta)\varphi = 0. \end{cases} \tag{2.2}$$

Lemma 2.1.

If (φ,η) *is a solitary wave of* (2.1) *for some* $c \in R$, *then* $c \neq 0$ *and* $\varphi(x) \neq c$ *for any* $x \in R$.

Proof.

From the definition of solitary waves and the embedding theorem we know that φ and η are both continuous. We first need to show that $c \neq 0$. Assume to the contrary that $c = 0$. Then (2.2) becomes

$$\begin{cases} \sigma\varphi\varphi_{xx} + \frac{\sigma}{2}\varphi_x^2 - \frac{1}{2}(\eta+1)^2 + A\varphi + \frac{1}{2} = 0, \\ (1+\eta)\varphi = 0. \end{cases} \tag{2.3}$$

Since $\eta \to 0$ as $|x| \to \infty$, the second equation of (2.3) implies $\varphi(x) = 0$ for $|x|$ sufficiently large. Let $\bar{x} = \max\{x : \varphi(x) \neq 0\}$.
Then $\varphi(x) \equiv 0$ on $[\bar{x}, \infty)$ and $\varphi(x) \not\equiv 0$ in $(\bar{x}-\delta, \bar{x})$ for any $\delta > 0$. Consider now the first equation of (2.3) on $[\bar{x}, \infty)$ we know that $\eta \equiv 0$ on $[\bar{x}, \infty)$. Then the continuity of η implies that there exists $\delta_1 > 0$ such that $1 + \eta(x) > 0$ on $(\bar{x} - \delta_1, \bar{x})$, combining which with the second equation of (2.3) leads to $\varphi(x) \equiv 0$ in $(\bar{x} - \delta_1, \bar{x})$. This is a contradiction. Therefore $c \neq 0$.

Next we have to show that $c-\varphi(x)\neq 0$. Suppose there is some $\bar{x}\in\mathbb{R}$ such that $c-\varphi(\bar{x})=0$. From the second equation of (2.2) we know $\varphi(\bar{x})=(c-\varphi(\bar{x}))\eta(\bar{x})=0$. So $c=(c-\varphi(\bar{x}))+\varphi(\bar{x})=0$, a contradiction to $c\neq 0$. Thus $\varphi(x)\neq c$.

Now we may rewrite (2.2) as

$$\begin{cases} (\sigma\varphi-c)\varphi_{xx}+\frac{\sigma}{2}\varphi_x^2-\frac{1}{2}(\eta+1)^2+A\varphi+\frac{1}{2}=0, \\ \eta=\frac{\varphi}{c-\varphi}. \end{cases} \tag{2.4}$$

Plugging $\eta=\frac{\varphi}{c-\varphi}$ into the first equation of (2.4) we obtain an equation for the unknown φ only

$$\begin{cases} -c\varphi_{xx}-\frac{1}{2}\frac{c^2}{(c-\varphi)^2}+A\varphi+\frac{1}{2}=0, & (\sigma=0), \\ \left(\varphi-\frac{c}{\sigma}\right)\varphi_{xx}+\frac{1}{2}\varphi_x^2-\frac{1}{2\sigma}\frac{c^2}{(c-\varphi)^2}+\frac{A}{\sigma}\varphi+\frac{1}{2\sigma}=0, & (\sigma\neq 0), \end{cases} \tag{2.5}$$

or equivalently,

$$\begin{cases} \varphi_{xx}=-\frac{1}{2}\frac{c}{(c-\varphi)^2}+\frac{A}{c}\varphi+\frac{1}{2c}, & (\sigma=0), \\ \left[(\varphi-\frac{c}{\sigma})^2\right]_{xx}=\varphi_x^2+\frac{1}{\sigma}\frac{c^2}{(c-\varphi)^2}-\frac{2A}{\sigma}\varphi-\frac{1}{\sigma}, & (\sigma\neq 0). \end{cases} \tag{2.6}$$

3. SOLITARY WAVES FOR Σ=0

Let us consider the first equation of (2.6)

$$\varphi_{xx} = -\frac{1}{2}\frac{c}{(c-\varphi)^2} + \frac{A}{c}\varphi + \frac{1}{2c}. \qquad (3.1)$$

By Lemma 2.1, we know $c \neq 0$ and $c-\varphi(x) \neq 0$. So the local regularity theory to elliptic Eq. (3.1) infers $\varphi(x) \in C^{\infty}(R)$ and so is η by the second equation of (2.4). Therefore all solitary waves are smooth in this case.

Multiplying both sides of (3.1) by $2\varphi_x$ and integrating on $(-\infty, x]$, it follows that

$$\varphi_x^2 = \frac{A}{c}\varphi^2 + \frac{1}{c}\varphi - \frac{c}{(c-\varphi)} + 1.$$

Then we obtain

$$\varphi_x^2 = \frac{\varphi^2(Ac - 1 - A\varphi)}{c(c-\varphi)}. \qquad (3.2)$$

So the numbers 0 and $c - \frac{1}{A}$ are the only possible extreme values of φ. We may rewrite (3.1) as the following system

$$\begin{cases} \varphi_x = v \\ v_x = -\frac{1}{2}\frac{c}{(c-\varphi)^2} + \frac{A}{c}\varphi + \frac{1}{2c}, \end{cases} \qquad (3.3)$$

with $c \neq 0$ and $c - \varphi \neq 0$. Notice that functions on the right-hand side of (3.3) are locally Lipschitz for φ and v at a neighborhood of one of the points $(0,0)$ and $(c - \frac{1}{A}, 0)$. To obtain nontrivial traveling waves, (3.3) implies $\varphi(x) \neq 0$ for $x \in R$. With the help of the assumption on $\varphi(x)\to 0$ as $|x|\to\infty$, we know from (3.2) that a necessary condition for the existence is $c \geq \frac{1}{A}$. But we may prove further that

Theorem 3.1.

Let $\sigma=0$. (2.1) *admits a solitary wave solution if and only if*

$$c > \frac{1}{A}. \tag{3.4}$$

All solitary waves are smooth in this case.

Proof.

We already discussed the regularity in the above. So we will just concentrate on the existence parts. If $c = \frac{1}{A}$, then (3.2) becomes

$$\varphi_x^2 = \frac{-A\varphi^3}{\frac{1}{A}\left(\frac{1}{A}-\varphi\right)} := G(\varphi). \tag{3.5}$$

From (3.5) we see that $\varphi(x)<0$ near $-\infty$ because of $A>0$. Since $\varphi(x)\to 0$ as $x\to-\infty$, there is some $\bar{x}$ sufficiently negative so that $\varphi(\bar{x}) = -\epsilon < 0$, with ϵ sufficiently small, and $\varphi_x(\bar{x}) < 0$. From the standard ODE theory, we can generate a unique local solution $\varphi(x)$ on $[\bar{x}-L, \bar{x}+L]$ for some $L>0$. Since $A>0$, we obtain

$$\left(\frac{-A\varphi^3}{\left(\frac{1}{A}-\varphi\right)}\right)' = \frac{\varphi^2(2A\varphi-3)}{\left(\frac{1}{A}-\varphi\right)^2} < 0, \quad \text{for } \varphi < 0. \tag{3.6}$$

Therefore $G(\varphi)$ decreases for $\varphi<0$. Since $\varphi_x(\bar{x}) < 0$, φ decreases near $\bar{x}$, so $G(\varphi)$ increases near $\bar{x}$. By (3.5) we know φ_x decreases near $\bar{x}$ and hence, both φ and φ_x decrease on $[\bar{x}-L, \bar{x}+L]$. Because $\sqrt{G(\varphi)}$ is locally Lipschitz in φ for $\varphi\leq 0$, we can easily continue the local solution to all of R and obtain that $\varphi\to-\infty$ as $x\to\infty$, which contradicts to be in $H^1(R)$. Thus there is no solitary wave when $c = \frac{1}{A}$. This completes the proof of Theorem 3.1.

4. SOLITARY WAVES FOR $\Sigma \neq 0$

In this case we consider the second equation of (2.6)

$$\left[\left(\varphi - \frac{c}{\sigma}\right)^2\right]_{xx} = \varphi_x^2 + \frac{c^2}{\sigma(c-\varphi)^2} - \frac{2A}{\sigma}\varphi - \frac{1}{\sigma}. \tag{4.1}$$

Inspired by the study of the traveling waves of the Camassa-Holm equation [37], we need the following Lemma 4.1 to deal with the regularity of the solitary waves.

Lemma 4.1.

Assume $\sigma \neq 0$ *and* (φ, η) *be a solitary wave of* (2.1). *Then*

$$\left(\varphi - \frac{c}{\sigma}\right)^k \in C^j\left(\mathbb{R} \setminus \varphi^{-1}\left(\frac{c}{\sigma}\right)\right), \quad k \geq 2^j. \tag{4.2}$$

Therefore

$$\varphi \in C^\infty\left(\mathbb{R} \setminus \varphi^{-1}\left(\frac{c}{\sigma}\right)\right). \tag{4.3}$$

Proof.

From Lemma 2.1 we know that $c \neq 0$ and $\varphi \neq c$ and thus φ satisfies (4.1). Let $v = \varphi - \frac{c}{\sigma}$ and denote

$$p(v) = -\frac{2A}{c}\left(v + \frac{c}{\sigma}\right) - \frac{1}{\sigma}.$$

So $p(v)$ is a polynomial in v. From the fact that $\varphi \neq c$ we see that

$$\frac{\sigma-1}{\sigma}c - v \neq 0. \tag{4.4}$$

Then v satisfies

$$(v^2)_{xx} = v_x^2 + p(v) + \frac{c^2}{\sigma}\left(\frac{\sigma-1}{\sigma}c - v\right)^{-2}.$$

From the assumption we see that $(v^2)_{xx} \in L^1_{loc}(\mathbb{R})$. Hence $(v^2)_x$ is absolutely continuous and hence

$v^2 \in C^1(\mathbb{R})$, and then $v \in C^1(\mathbb{R} \setminus v^{-1}(0))$.

By (4.4) and $v + \frac{c}{\sigma} \in H^1(\mathbb{R}) \subset C(\mathbb{R})$, we know

$$\left(\frac{\sigma-1}{\sigma}c - v\right)^{-2} \in C(\mathbb{R}) \cap C^1(\mathbb{R} \setminus v^{-1}(0)).$$

Moreover,

$$\begin{aligned}(v^k)_{xx} &= (kv^{k-1}v_x)_x = \frac{k}{2}(v^{k-2}(v^2)_x)_x = k(k-2)v^{k-2}v_x^2 + \frac{k}{2}v^{k-2}(v^2)_{xx} \\ &= k(k-2)v^{k-2}v_x^2 + \frac{k}{2}v^{k-2}\left[v_x^2 + p(v) + \frac{c^2}{\sigma}\left(\frac{\sigma-1}{\sigma}c - v\right)^{-2}\right] \\ &= k\left(k - \frac{3}{2}\right)v^{k-2}v_x^2 + \frac{k}{2}v^{k-2}p(v) + \frac{kc^2}{2\sigma}v^{k-2}\left(\frac{\sigma-1}{\sigma}c - v\right)^{-2}.\end{aligned} \tag{4.5}$$

For k = 3, the right-hand side of (4.5) is in $L^1_{loc}(\mathbb{R})$. We deduce that

$$v^3 \in C^1(\mathbb{R}).$$

For k ≥ 4, we notice that (4.5) implies

$$(v^k)_{xx} = \frac{k}{4}\left(k - \frac{3}{2}\right) v^{k-4}[(v^2)_x]^2 + \frac{k}{2} v^{k-2} p(v) + \frac{kc^2}{2\sigma} v^{k-2} \left(\frac{\sigma - 1}{\sigma} c - v\right)^{-2}.$$

Since $v^2 \in C^1(\mathbb{R})$, it follows that

$$v^k \in C^2(\mathbb{R}), \quad k \geq 4.$$

For k ≥ 8, we see from the above that $v^4, v^{k-4}, v^{k-2}, v^{k-2}p(v) \in C^2(\mathbb{R})$, and $v^{k-2}\left(\frac{\sigma-1}{\sigma}c - v\right)^{-2} \in C^2(\mathbb{R} \setminus v^{-1}(0))$. Moreover, we have

$$v^{k-2} v_x^2 = \frac{1}{4}(v^4)_x \frac{1}{k-4}(v^{k-4})_x \in C^1(\mathbb{R}).$$

We conclude from (4.5) that

$$v^k \in C^3(\mathbb{R} \setminus v^{-1}(0)), \quad k \geq 8.$$

Using these arguments to higher values of k we prove that $v^k \in C^j(R \setminus v^{-1}(0))$ for $k \geq 2^j$, and hence (4.2). This completes the proof of Lemma 4.1.

Set $x_0 = \min\{x : \varphi(x) = \frac{c}{\sigma}\}$ (if $\varphi \neq \frac{c}{\sigma}$ for all x then let $x_0 = +\infty$). Then we know $x_0 \leq +\infty$. From Lemma 4.1, a solitary wave φ is smooth on $(-\infty, x_0)$ and hence the second equation of (2.6) holds pointwise on $(-\infty, x_0)$. Therefore we may multiply both sides of the second equation of (2.6) by $2\varphi_x$ and integrate on $(-\infty, x]$ for $x < x_0$ to get

$$\varphi_x^2 = \frac{\varphi^2(Ac - 1 - A\varphi)}{(c - \varphi)(c - \sigma\varphi)} := F(\varphi). \tag{4.6}$$

Using the idea as introduced in [37] it gives us the following conclusions.

(i) Assume $F(\varphi)$ has a simple zero at $m = c - \frac{1}{A}$ so that $F(m) = 0$ and $F'(m) \neq 0$. The solution φ of (4.6) satisfies

$$\varphi_x^2 = (\varphi - m)F'(m) + O((\varphi - m)^2) \quad \text{as } x \to m,$$

where $f = O(g)$ as $x \to a$ means that $\left|\frac{f(x)}{g(x)}\right|$ is bounded in some interval $[a - \epsilon, a + \epsilon]$ with $\epsilon > 0$. Therefore

$$\varphi(x) = m + \frac{1}{4}(x - \xi)^2 F'(m) + O((x - \xi)^4) \quad \text{as } x \to \xi, \tag{4.7}$$

where $\varphi(\xi) = m$.

(ii) If $F(\varphi)$ has a double zero at $m = 0$, so that $F'(0) = 0, F''(0) > 0$, then

$$\varphi_x^2 = \varphi^2 F''(0) + O(\varphi^3) \quad \text{as } \varphi \to 0.$$

We obtain

$$\varphi(x) \sim \alpha \exp\left(-x\sqrt{F''(0)}\right) \quad \text{as } x \to \infty \tag{4.8}$$

for some constant α. Thus $\varphi \to 0$ exponentially as $x \to \infty$.

(iii) If φ approaches a simple pole $\varphi(x_0) = \frac{c}{\sigma}$ of $F(\varphi)$ (when $\sigma \neq 1$), then

$$\varphi(x) - \frac{c}{\sigma} = \beta|x - x_0|^{2/3} + O((x - x_0)^{4/3}) \quad \text{as } x \to x_0, \tag{4.9}$$

$$\varphi_x = \begin{cases} \frac{2}{3}\beta|x - x_0|^{-1/3} + O((x - x_0)^{1/3}) & \text{as } x \to x_0, \\ -\frac{2}{3}\beta|x - x_0|^{-1/3} + O((x - x_0)^{1/3}) & \text{as } x \to x_0 \end{cases} \tag{4.10}$$

for some constant β. In particular, when F has a pole, the solution φ has a cusp.

(iv) If the change of direction: $\varphi_x \mapsto -\varphi_x$ according to (4.6), then peaked solitary waves occur.

In view of this discussion (i)–(iv), we give the following theorem on the existence of solitary waves of (2.1) for $\sigma \neq 0$.

Theorem 4.1.

Let $\sigma \neq 0$. Any solitary wave of (2.1) for $\sigma \neq 0$ belongs to one of the following categories.

(i) *For $0 < \sigma \leq 1$:*

- *A solitary wave (φ, η) of (2.1) exists if and only if condition (3.4) holds.*
- *If $c > \frac{1}{A}$, then there exist a smooth solitary wave $\varphi > 0$ with $\max_{x \in \mathbb{R}} \varphi(x) = c - \frac{1}{A}$.*

(ii) *For $\sigma < 0$:*

- *If* $c > \frac{1}{A}$, *then there exist a smooth solitary wave* $\varphi > 0$ *with* $\max_{x\in\mathbb{R}}\varphi(x) = c - \frac{1}{A}$, *and an anticusped solitary wave (the solution profile has a cusp pointing downward)* $\varphi < 0$ *with* $\min_{x\in\mathbb{R}}\varphi(x) = \frac{c}{\sigma}$.

- *If* $c = \frac{1}{A}$, *then there exist an anticusped solitary wave* $\varphi < 0$ *with* $\min_{x\in\mathbb{R}}\varphi(x) = \frac{c}{\sigma}$.

(iii)*For* $\sigma > 1$*:*

- *A solitary wave exists if and only if* c *satisfies* (3.4). *If* $c > \frac{1}{A}$, *then* $\varphi > 0$.

- *If* $\frac{1}{A} < c < \frac{\sigma}{(\sigma-1)A}$, *then the solitary waves are smooth and unique up to translation with* $\max_{x\in\mathbb{R}}\varphi(x) = c - \frac{1}{A}$.

- *If* $c = \frac{\sigma}{(\sigma-1)A}$, *then the solitary wave is peaked with* $\max_{x\in\mathbb{R}}\varphi(x) = c - \frac{1}{A} = \frac{c}{\sigma}$.

If $c > \frac{\sigma}{(\sigma-1)A}$, *then the solitary waves are cusped with* $\max_{x\in\mathbb{R}}\varphi(x) = \frac{c}{\sigma}$.

Moreover, each kind of the above solitary waves is unique and even up to translations. When $c > \frac{1}{A}$, *all solitary waves decay exponentially to zero at infinity.*

Proof.

First from (4.6) and the assumption $\varphi(x) \to 0$ as $|x| \to \infty$ we see that a necessary condition for the existence of a solitary wave is that $c \geq \frac{1}{A}$. If $c = \frac{1}{A}$ then (4.6) becomes

$$\varphi_x^2 = \frac{-A\varphi^3}{\left(\frac{1}{A} - \varphi\right)\left(\frac{1}{A} - \sigma\varphi\right)} := F_1(\varphi). \tag{4.11}$$

Hence we know that $\varphi(x)<0$ near $-\infty$. Similarly as in the proof of Theorem 3.1, we can find some $\bar{x}$ sufficiently large negative with $\varphi(\bar{x}) = -\epsilon < 0$ and $\varphi_x(\bar{x}) < 0$, and we can establish a unique local solution $\varphi(x)$ on $[\bar{x} - L, \bar{x} + L]$ for some $L>0$.

If $\sigma<0$, we know that $\frac{1}{\frac{1}{A}-\sigma\varphi}$ is decreasing when $\varphi<0$. With the help of (3.5) we see that $F_1(\varphi)$ decreases for $\varphi<0$. Since we have $\varphi_x(\bar{x}) < 0, \varphi$ decreases near $\bar{x}$. So $F_1(\varphi)$ increases near $\bar{x}$. Therefore from (4.11), φ_x decreases near $\bar{x}$, and then both φ and φ_x decrease on $[\bar{x} - L, \bar{x} + L]$. Since $\sqrt{F_1(\varphi)}$ is locally Lipschitz in φ for $\frac{1}{\sigma A} < \varphi \leq 0$, we can easily continue the local solution to $(-\infty, \bar{x} - L]$ with $\varphi(x)\to 0$ as $x\to-\infty$. On $[\bar{x} + L, +\infty)$, we can solve the initial problem

$$\begin{cases} \psi_x = -\sqrt{F_1(\psi)}, \\ \psi(\bar{x} + L) = \varphi(\bar{x} + L) \end{cases}$$

all the way until $\psi = \frac{1}{\sigma A}$, which is a simple pole of $F_1(\psi)$. Using and, we can establish an anticusped solution with a cusp singularity at $\varphi = \frac{1}{\sigma A} = \frac{c}{\sigma}$.

If $\sigma>0$, a direct computation shows that

$$F_1'(\varphi) < 0, \quad \text{for } \varphi < 0.$$

Therefore the same argument implies that $\varphi(x)\to-\infty$ as $x\to+\infty$, which fails to be in $H^1(R)$. Hence there is no solitary wave in this case.

Now we consider $c > \frac{1}{A}$. From (4.6) we see that φ cannot oscillate around zero near infinity. Let us consider the following two cases:

Case 1. $\varphi(x)>0$ near $-\infty$. Then there is some $\bar{x}$ sufficiently large negative so that $\varphi(\bar{x}) = \epsilon > 0$, with ϵ sufficiently small, and $\varphi_x(\bar{x}) > 0$.

When $0 < \sigma \leq 1$, $\sqrt{F(\varphi)}$ is locally Lipschitz in φ for $0 \leq \varphi \leq c - \frac{1}{A}$. Hence there is a local solution to

$$\begin{cases} \varphi_x = \sqrt{F(\varphi)}, \\ \varphi(\bar{x}) = \epsilon \end{cases}$$

on $[\bar{x} - L, \bar{x} + L]$ for some L>0. Therefore from (4.7) and (4.8) we obtain a smooth solitary wave solution with maximum height $\varphi = c - \frac{1}{A}$ and an exponential decay to zero at infinity

$$\varphi(x) = O\left(\exp\left(-\frac{\sqrt{2(Ac-1)}}{c}|x|\right)\right) \quad \text{as } |x| \to \infty. \tag{4.12}$$

When $\sigma > 1$, $\sqrt{F(\varphi)}$ is locally Lipschitz in φ for $0 \leq \varphi < \frac{c}{\sigma}$. Using the previous arguments, if $c - \frac{1}{A} < \frac{c}{\sigma}$ (i.e., $c < \frac{\sigma}{(\sigma-1)A}$), then we obtain smooth solitary waves with exponential decay.

If $c - \frac{1}{A} = \frac{c}{\sigma}$ (i.e., $c = \frac{\sigma}{(\sigma-1)A}$) then the smooth solution can be established until $\varphi = c - \frac{1}{A} = \frac{c}{\sigma}$. However, it can make a sudden turn at $\varphi = c - \frac{1}{A} = \frac{c}{\sigma}$ and so give rise to a peak. Sinceφ=0 is still a double zero of F(φ), we still have the exponential decay here.

Lastly if $c - \frac{1}{A} > \frac{c}{\sigma}$ (i.e., $c > \frac{\sigma}{(\sigma-1)A}$) then $\varphi = \frac{c}{\sigma}$ becomes a pole of F(φ). Hence from and we see that we obtain a solitary wave with a cusp at $\varphi = \frac{c}{\sigma}$ and decaying exponentially.

Case 2. φ(x)<0 near −∞. In this case we solve

$$\begin{cases} \varphi_x = -\sqrt{F(\varphi)}, \\ \varphi(\bar{x}) = -\epsilon \end{cases}$$

for some $\bar{x}$ sufficiently large negative and ϵ>0 sufficiently small.

When $\sigma>0$ we see that $F'(\varphi)<0$ for $\varphi<0$. Thus there is no solitary wave in this case.

When $\sigma<0$, we know $\varphi=\frac{c}{\sigma}<0$ is a pole of $F(\varphi)$. Hence from the previous arguments, we obtain an anticusped solitary wave with $\min_{x\in\mathbb{R}}\varphi(x)=\frac{c}{\sigma}$, which decays exponentially.

Finally, from the standard ODE theory and the fact that the second equation of (2.5) is invariant under the transformations $x\mapsto x+k$ for any constant k, and $x\mapsto -x$, we conclude that the solitary waves obtained above are unique and even up to translations. This completes the proof of Theorem 4.1. □

Next we try to find an implicit formula for the peaked solitary waves. Consider the case $\sigma>1$ and $c>\frac{1}{A}$. Then from the Theorem 4.1 we see that peaked solitary waves exist only when $c=\frac{\sigma}{(\sigma-1)A}$. In this case we have

$$\varphi_x^2=\frac{A\varphi^2}{\sigma(c-\varphi)}.$$

Since φ is positive, even with respect to some $\bar{x}$ and decreasing on $(\bar{x},\infty)$, so for $x>\bar{x}$ we have

$$\varphi_x=-\varphi\sqrt{\frac{A}{\sigma(c-\varphi)}}.$$

Thus from the separation of variables we obtain

$$-(x-\bar{x})=\int_{c-\frac{1}{A}}^{\varphi}\frac{1}{t\sqrt{\frac{A}{\sigma(c-t)}}}dt.$$

Letting $w=\frac{A}{\sigma(c-t)}$, the above becomes

$$
\begin{aligned}
-(x-\bar{x}) &= \int_{\frac{A^2}{\sigma}}^{\frac{A}{\sigma(c-\varphi)}} \frac{\frac{A}{\sigma}}{(cw-\frac{A}{\sigma})w^{3/2}}dw = \int_{\frac{A^2}{\sigma}}^{\frac{A}{\sigma(c-\varphi)}} \frac{1}{\sqrt{w}}\left(\frac{c}{cw-\frac{A}{\sigma}}-\frac{1}{w}\right)dw \\
&= \int_{\frac{A^2}{\sigma}}^{\frac{A}{\sigma(c-\varphi)}} \left\{\sqrt{\frac{c\sigma}{A}}\left(\frac{\frac{1}{2}\sqrt{\frac{c}{w}}}{\sqrt{cw}-\sqrt{\frac{A}{\sigma}}}-\frac{\frac{1}{2}\sqrt{\frac{c}{w}}}{\sqrt{cw}+\sqrt{\frac{A}{\sigma}}}\right)+\frac{1}{w^{3/2}}\right\}dw \\
&= \left(\sqrt{\frac{c\sigma}{A}}\ln\left|\frac{\sqrt{cw}-\sqrt{\frac{A}{\sigma}}}{\sqrt{cw}+\sqrt{\frac{A}{\sigma}}}\right|\right)\Bigg|_{\frac{A^2}{\sigma}}^{\frac{A}{\sigma(c-\varphi)}} + \frac{2\sqrt{\sigma}}{A} - \frac{2\sqrt{\sigma(c-\varphi)}}{\sqrt{A}}.
\end{aligned}
$$

Therefore we obtain an implicit formula for the peaked solitary waves.

equation(4.13)

$$
-|x-\bar{x}| = \left(\sqrt{\frac{c\sigma}{A}}\ln\left|\frac{\sqrt{cw}-\sqrt{\frac{A}{\sigma}}}{\sqrt{cw}+\sqrt{\frac{A}{\sigma}}}\right|\right)\Bigg|_{\frac{A^2}{\sigma}}^{\frac{A}{\sigma(c-\varphi)}} + \frac{2\sqrt{\sigma}}{A} - \frac{2\sqrt{\sigma(c-\varphi)}}{\sqrt{A}}.
$$

A peaked solitary wave profile of (4.13) corresponding to $\bar{x}=0, A=2, \sigma=2$, and c=1 is plotted in Fig. 1.

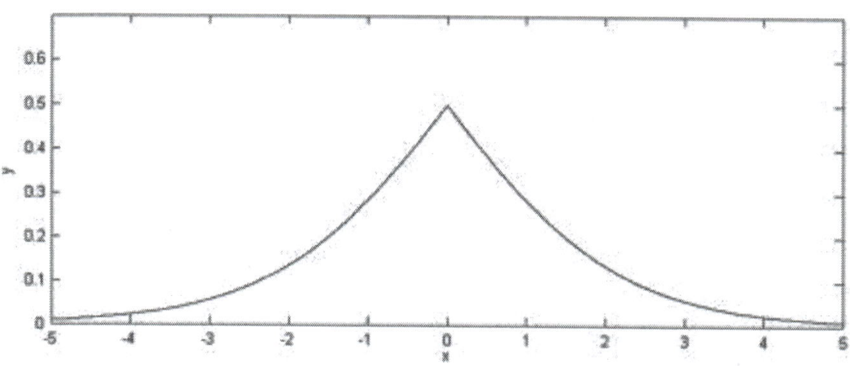

Figure 1. A peaked solitary wave profile y=φ(x) with $\bar{x}=0, A=2, \sigma=2$, and c=1.

ACKNOWLEDGMENTS

This work is partially supported by the NSF grant DMS-0906099 and the NHARP grant-003599-0001-2009.

REFERENCES

1. R.M. Chen, Y. Liu, Wave breaking and global existence for a generalized two-component Camassa–Holm system, Int. Math. Res. Not. 6 (2011) 1381–1416.
2. R. Ivanov, Two-component integrable systems modelling shallow water waves: the constant vorticity case, Wave Motion 46 (2009) 389–396.
3. A. Constantin, J. Escher, Analyticity of periodic traveling free surface water waves with vorticity, Ann. of Math. 173 (2011) 559–568.
4. W. Strauss, Steady water waves, Bull. Amer. Math. Soc. 47 (2010) 671–694.
5. J. Escher, Non-metric two-component Euler equation on the circle, Monatsh. Math. http://dx.doi.org/10.1007/s00605-011-0323-3.
6. J. Lenells, O. Lechtenfeld, On the N = 2 supersymmetric Camassa–Holm and Hunter–Saxton systems, J. Math. Phys. 50 (2009) 1–17.
7. J. Liu, Z. Yin, Blow-up phenomena and global existence for a periodic two-component Hunter–Saxton system, Preprint.
8. H. Wu, M. Wunsch, Global existence for the generalized two-component Hunter–Saxton system, J. Math. Fluid Mech.
9. M. Wunsch, On the Hunter–Saxton system, Discrete Contin. Dyn. Syst. 12 (2009) 647–656.
10. M. Wunsch, The generalized Hunter–Saxton system, SIAM J. Math. Anal. 42 (2010) 1286–1304.

11. M. Wunsch, Weak geodesic flow on a semi-direct product and global solutions to the periodic Hunter–Saxton system, Nonlinear Anal. 74 (2011) 4951–4960.

12. A. Constantin, R.I. Ivanov, On an integrable two-component Camassa–Holm shallow water system, Phys. Lett. A 372 (2008) 7129–7132.

13. J.K. Hunter, R. Saxton, Dynamics of director fields, SIAM J. Appl. Math. 51 (1991) 1498–1521.

14. R. Beals, D.H. Sattinger, J. Szmigielski, Inverse scattering solutions of the Hunter–Saxton equation, Appl. Anal. 78 (3&4) (2001) 255–269.

15. A. Bressan, A. Constantin, Global solutions of the Hunter–Saxton equation, SIAM J. Math. Anal. 37 (3) (2005) 996–1026.

16. A. Bressan, H. Holden, X. Raynaud, Lipschitz metric for the Hunter–Saxton equation, J. Math. Pures Appl. 94 (2010) 68–92.

17. F. Tiğlay, The periodic Cauchy problem of the modified Hunter–Saxton equation, J. Evol. Equ. 5 (4) (2005) 509–527.

18. Z. Yin, On the structure of solutions to the periodic Hunter–Saxton equation, SIAM J. Math. Anal. 36 (2004) 272–283.

19. H.H. Dai, M. Pavlov, Transformations for the Camassa–Holm equation, its high-frequency limit and the Sinh–Gordon equation, J. Phys. Soc. Japan 67 (1998) 3655–3657.

20. J.K. Hunter, Y. Zheng, On a completely integrable hyperbolic variational equation, Physica D 79 (1994) 361–386.

21. R. Camassa, D.D. Holm, An integrable shallow water equation with peaked solitons, Phys. Rev. Lett. 71 (11) (1993) 1661–1664.

22. A. Constantin, D. Lannes, The hydrodynamical relevance of the Camassa–Holm and Degasperis–Procesi equations, Arch. Ration. Mech. Anal. 192 (2009) 165–186.

23. R.S. Johnson, Camassa–Holm, Korteweg–de Vries and related models for water waves, J. Fluid Mech. 455 (2002) 63–82.

24. D. Ionescu-Kruse, Variational derivation of the Camassa–Holm shallow water equation, J. Nonlinear Math. Phys. 14 (2007) 303–312.

25. P. Olver, P. Rosenau, Tri-Hamiltonian duality between solitons and solitary wave solutions having compact support, Phys. Rev. E 53 (1996) 1900–1906.

26. B. Khesin, J. Lenells, G. Misiolek, Generalized Hunter–Saxton equation and the geometry of the group of circle diffeomorphisms, Math. Ann. 342 (2008) 617–656.

27. J. Lenells, Periodic solitons of an equation for short capillary-gravity waves, J. Math. Anal. Appl. 352 (2009) 964–966.

28. M.S. Alber, R. Camassa, D.D. Holm, J.E. Marsden, The geometry of peaked solitons and billiard solutions of a class of integrable PDE's, Lett. Math. Phys. 32 (1994) 137–151.

29. A. Constantin, The trajectories of particles in Stokes waves, Invent. Math. 166 (2006) 523–535.

30. A. Constantin, J. Escher, Particle trajectories in solitary water waves, Bull. Amer. Math. Soc. 44 (2007) 423–431.

31. J.F. Toland, Stokes waves, Topol. Methods Nonlinear Anal. 7 (1996) 1–48.

32. A. Constantin, W. Strauss, Stability of peakons, Comm. Pure Appl. Math. 53 (2000) 603–610.

33. K. El Dika, L. Molinet, Exponential decay of H1-localized solutions and stability of the train of N solitary waves for the Camassa–Holm equation, Philos.

34. Trans. Roy. Soc. London A 365 (2007) 2313–2331.

35. J. Lenells, A variational approach to the stability of periodic peakons, J. Nonlinear Math. Phys. 11 (2004) 151–163.

36. A. Constantin, W. Strauss, Stability of the Camassa–Holm solitons, J. Nonlinear Sci. 12 (2002) 415–422.

37. B. Moon, Y. Liu, Wave breaking and global existence for the generalized periodic two-component Hunter–Saxton system, J. Differential Equations 253 (2012) 319–355.

38. J. Lenells, Traveling wave solutions of the Camassa–Holm equation, J. Differential Equations 217 (2005) 393–430.

3

CHAPTER

Relative Equilibria in the Spherical, Finite Density Three-Body Problem

D. J. Scheeres

Department of Aerospace Engineering Sciences, The University of Colorado at Boulder

ABSTRACT

The relative equilibria for the spherical, finite density three-body problem are identified. Specifically, there are 28 distinct relative equilibria in this problem which include the classical five relative equilibria for the point-mass three-body problem. None of the identified relative equilibria exist or are stable over all values of angular momentum. The stability and bifurcation pathways of these relative equilibria are mapped out as the angular momentum of the system is increased. This is done under the assumption that they have equal and constant densities and that the entire system rotates about its maximum moment of inertia. The transition to finite density greatly increases the number of relative equilibria in the three-body problem and ensures that minimum energy configurations exist for all values of angular momentum.

Keywords: 3-body problem Relative equilibria Granular mechanics

1 INTRODUCTION

The three-body problem is one of the most fundamental and well-studied problems in celestial mechanics. A key result for this problem is that there exist only five relative equilibria and that these exist for all levels of angular momentum (Euler 1767; Lagrange 1772). The properties of these special solutions have been deeply studied and have motivated significant research in mechanics and dynamics. A hallmark of the classical problem is that the bodies are considered to be point masses, with no restrictions on how close they can come to each other. A recent variation of this problem has been posed that removes this one restriction (Scheeres 2012) and supposes that these bodies can be rigid bodies with finite density, and hence have limits on their proximity. Such "full-body" systems inherit the fundamental symmetries of the N-body problem (Scheeres 2002); however, they also demand that the rotational angular momentum, energy and dynamics of these rigid bodies be incorporated in the theory as well.

This paper studies the spherical three-body problem under the assumption that the bodies are rigid and have finite density, and thus the separation between the bodies is constrained to be positive. This one change completely alters the character of the problem, and while the traditional Euler and Lagrange solutions still exist for large enough angular momentum values, a full 23 additional relative equilibria emerge from the analysis at all values of angular momentum, with a complex and rich bifurcation scheme.

The celestial mechanics of bodies with finite density and fixed shape can have dynamical evolution and relative equilibria that are quite distinct from that found in the classical Newtonian point-mass N-body problem. These differences were previously explored in Scheeres (2012) where several results were proven for the so-called full-body problem, in which the individual bodies are treated as rigid bodies with finite densities. Specifically, it was shown that, in opposition to the point-mass N-body problem, the full-body problem will always have a minimum energy configuration. Further, the number and variety of relative equilibria for that problem are greatly enhanced and now include configurations where the bodies can rest on each other and configurations where different collections of resting bodies orbit each other, as well as the classical central configurations. One important aspect of this problem is that the existence and stability of configurations become a function of the total angular momentum of the system, a dependance that does not exist for the classical point-mass *N*-body problem.

This paper studies the relative equilibria of one particular problem in the full-body problem (FBP) to completeness. Specifically, all relative equilibria of the planar spherical full three-body problem, which consists of three spheres of equal density but arbitrary size, located in the plane perpendicular to the angular momentum. The explicit methodology used was developed in Scheeres (2012, in press), and is fundamentally based on analysis of the amended potential as developed by Smale (1970a, (1970b) and motivated by observations from Arnold et al. (1988). The main theorem is stated and described at first, the problem is technically defined, then several results used to make the proof are listed, and finally all the detailed computations for the proof are given. Following the proof, a summary of the proof is provided, indicating how it establishes the theorem.

A main application of this result is to identify the stable states that can be physically achieved by a collection of self-gravitating bodies that can sustain contact. This situation happens in solar system dynamics when considering the physical nature of rubble pile asteroids (Fujiwara et al. 2006), although there the number of individual grains can be quite large. Recent observations of comets, however, also show that they can be comprised of a few larger components that rest on each other and, given their changing spin rates due to outgassing, also mimic a system with changing angular momentum (Massironi et al. 2015). A specific motivation from this current analysis would seek out natural situations that mimic the stable members of these configurations. As the existence of some of these stable configurations are somewhat unexpected, finding such configurations in nature would be especially interesting.

Additional avenues for exploration would be to expand the current analyses to study the dynamics of finite bodies as they interact with each other gravitationally and through impact. This sort of approach has been followed in the planetary sciences community in the study of rubble pile asteroids (Richardson et al. 2005; Sánchez and Scheeres 2012). The current results can motivate the formation of stable shapes as a function of body morphology and total angular momentum. In addition, the presence of finite densities and minimum energy configurations also enables the rigorous computation of energy limits for the Hill stability in the full *N*-body problem (Scheeres 2014, 2015), an avenue of further investigation for the current problem.

2 MAIN RESULT

Theorem 1

In the spherical full three-body problem, there exist a total of 28 distinct relative equilibrium configurations. No single class of relative equilibria exists or is energetically stable for all angular momentum; however, at every value of angular momentum, there exists at least 1 energetically stable relative equilibria. The pattern of relative equilibria can be fully represented in a bifurcation chart as the total angular momentum of the system varies from 0 to $\infty\infty$.

The 28 different relative equilibria can be delineated in a few different ways. Figure 1 shows these relative equilibria separated into seven different classes, each with one-to-six distinct configurations. The figure shows 20 distinct configurations, with eight of them having an alternate ordering not shown in the figure.

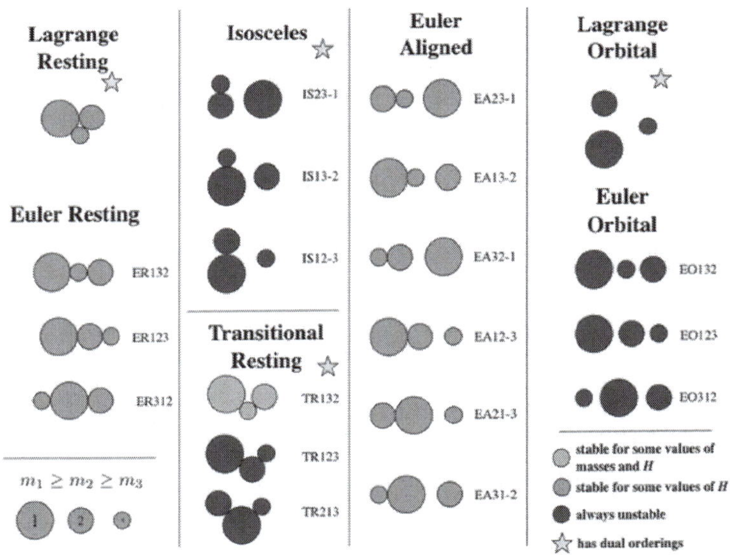

Figure 1. Examples of the 20 different equilibrium configurations are shown within different classes. *Colors* indicate whether or not the configuration can be stable for a range of angular momentum (*green*), for some special combinations of mass ratio and angular momentum (*yellow*), or if they are unstable for all angular momentum values and mass ratios (*red*). *Stars* indicate when a reordering of the masses provides another relative equilibrium. The detailed naming convention for the configurations is also introduced (Color figure online)

Figures 2, 3, 4 and 5 show the detailed sequences of bifurcations that occur as the total angular momentum of the system is increased. These diagrams are qualitative, but the sequence of bifurcations in the specific connected pathways are accurate and will be derived in the course of the proof. Note that the sequence of bifurcations keeps some of the configurations separated from each other. In other cases the sequence of bifurcations will change as the relative masses of the bodies are changed, in these cases multiple types of sequences are shown. Note in Fig. 3 that this particular sequence has at least one stable configuration for all values of angular momentum, while none of the others has a stable configuration in the vicinity of H=0H=0. Also note that the sequences shown in Figs. 2, 3 and 4 all have two stable configurations for large *H*.

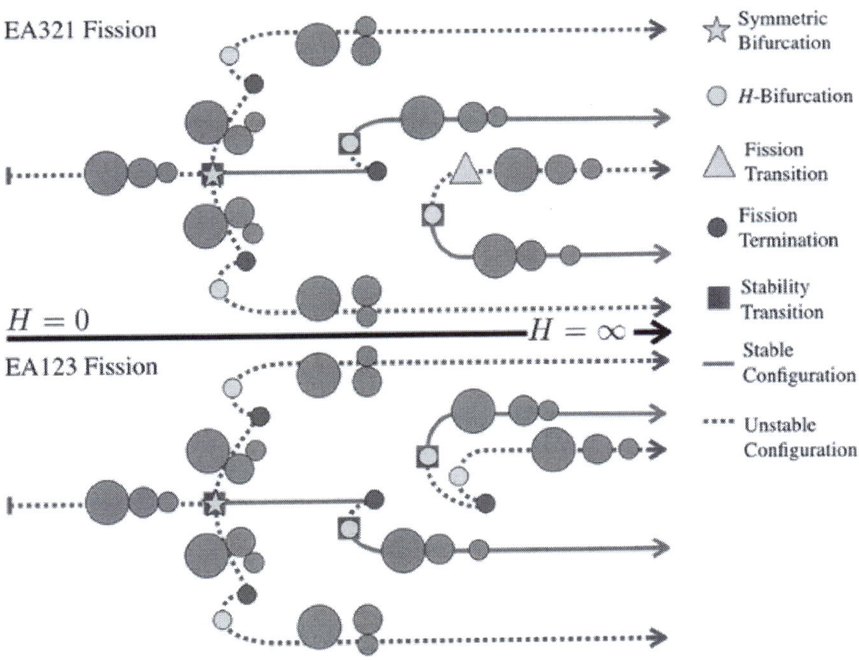

Figure 2. Bifurcation diagram showing the possible branches relating to the ER123 configuration. Depending on the relative mass values two different bifurcation pathways exist. Within each pathway, the manner in which the EO configuration appears can shift between the two pathways shown

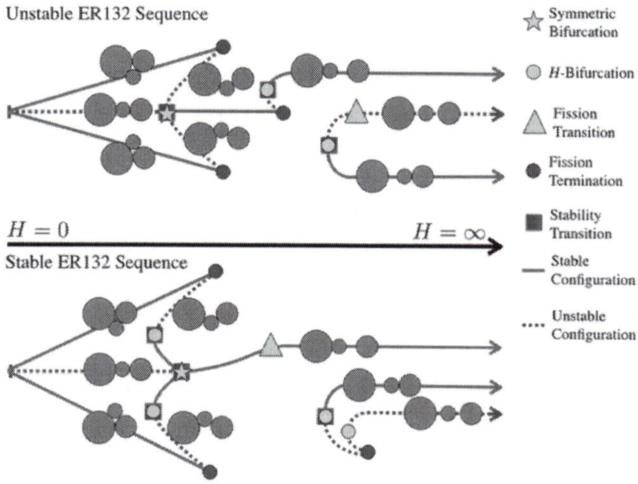

Figure 3. Bifurcation diagram showing the possible branches relating to the ER132 configuration. ER132 links to the LR configuration and always fissions into the same EA13-2 configuration. Depending on the relative mass values there can be a range of stable TR132 configurations and stable fission transitions to the EA13-2 relative equilibrium. Within each pathway, the manner in which the EO configuration appears can also shift between the two pathways shown. The two different sequence need not have the same pattern of TR and EO bifurcations

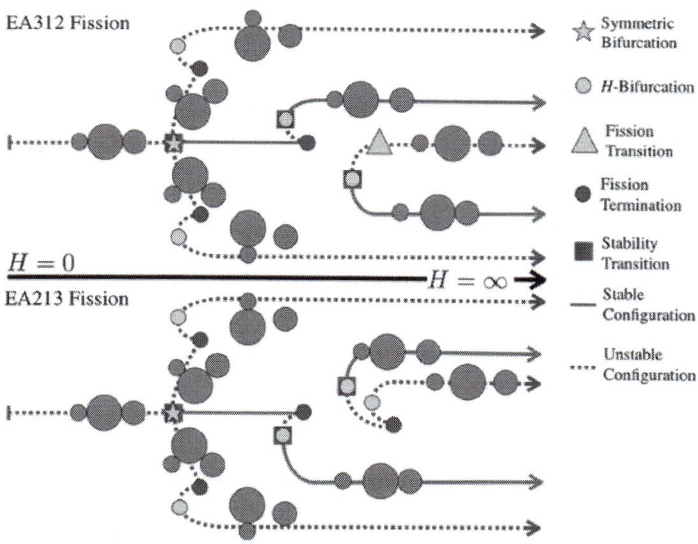

Figure 4. Bifurcation diagram showing the possible branches relating to the ER213 configuration. Depending on the relative mass values two different bifurcation pathways exist. Within each pathway, the manner in which the EO configuration appears can also shift between the two pathways shown

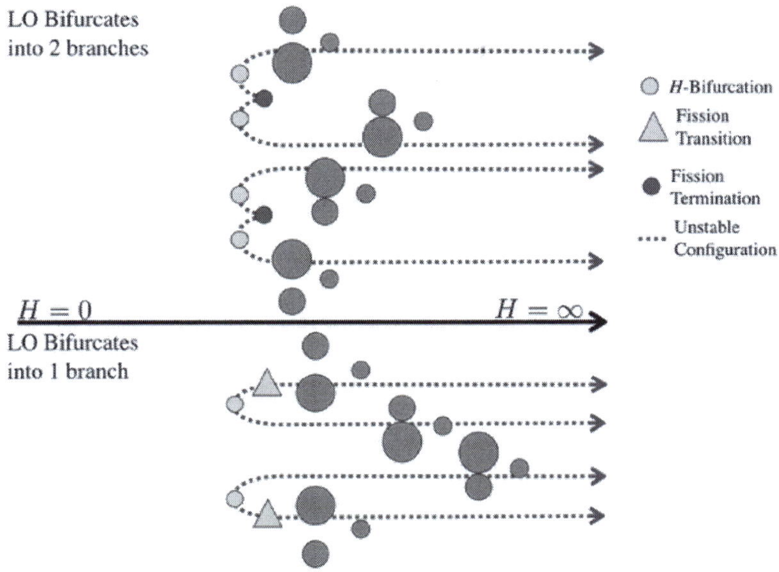

Figure 5. Bifurcation diagram showing the pathways to the LO configuration

3 PROBLEM STATEMENT

3.1 The Full-Body Problem

A full-body problem is defined as a set of *N* rigid bodies that gravitationally attract each other and which have a finite density mass distribution, meaning that there are specific limits on how close they can come to each other (Scheeres 2002, 2012, in press). The description of such a system can be directly incorporated into a Lagrangian framework where the coordinates $\{Q_i; i = 1, 2, \ldots, 6N\}$ denote the absolute Cartesian coordinates of the bodies and the Euler angles that orient the bodies in space. The rigid body constraints place restrictions of the form $|Q_i - Q_j| \geq D_{ij}(\mathrm{Q})$ on the system. The dynamics of the system can be described by a total Kinetic Energy and Gravitational Potential Energy, and as it is an isolated system will conserve its total angular momentum, denoted as ***H***, and can conserve its total energy, denoted as *E*, if internal dissipative forces are excluded.

The linear momentum can be removed to reduce the system to 6(*N*–*1*) coordinates that are purely relative to each other and an additional three degrees of freedom that orient the entire system with respect to inertial space. Denote the relative coordinates as $\mathrm{q} = \{q_i; i = 1, 2, \ldots 6(N-1)\}$, noting that these can always be transformed to locally reformulate the constraints as $q_i \geq D_i(\mathrm{q})$.

Thus, for the full three-body problem we have 12 relative degrees of freedom between the three bodies. Of these only three are required to specify the relative positions of the bodies. The additional nine correspond to each of the bodies having three degrees of freedom to orient themselves relative to the position configuration of the bodies. As we take the three bodies to be spheres, their relative orientation is not tracked, although we must still account for their rotational angular momentum and kinetic energy. Thus, for our purposes, our problem can be specified with only three degrees of freedom, plus the overall orientation of the system with respect to inertial space.

Before continuing we define the amended potential, which plays a fundamental role in the following.

Definition 1

Amended Potential The Amended Potential is defined as the function $\mathcal{E} = \frac{H^2}{2I_H} + \mathcal{U}$ where *H* is the total angular momentum of the system, ***IH*** is the moment of inertia of the total system taken about a principal axis of the system, in general about the rotation axis ***H*** which points in the direction of the total angular momentum vector, and UU is the gravitational potential energy of the system. The terms IHIH and UU are functions only of the relative positions and attitudes of the bodies, and for IHIH their orientation relative to HH^HH^. The gradients of the Amended Potential with respect to the degrees of freedom equal the force exerted on that degree of freedom when at an equilibrium or resting configuration (Arnold et al. 1988, pp. 66–67).

3.2 Spherical Full Three-Body Problem Statement

Consider three bodies, Bi,i=1,2,3Bi,i=1,2,3, each of which is a sphere of radius RiRi and, for convenience, assumed to have a common density ρρ.

The positions of these bodies can be denoted in R3R3 by Cartesian position vectors rrirri. The relative positions of these bodies are denoted as and have the fundamental rigid body constraint $|r_{ij}| \geq (R_i + R_j)$ for $i \neq j$. This lower bound, due to the bodies having finite density, is what enables resting equilibria to occur. Each of the spheres can carry angular momentum in their spin rate, although due to their symmetry the specific orientation of these spheres are arbitrary in any frame. Thus, the internal relative configuration space of the system, qqqq, is completely specified by only three quantities

$$q = \{r_{12}, r_{23}, r_{31} \mid r_{ij} \geq (R_i + R_j) \;\&\; |r_{ij} - r_{jk}| \leq r_{ki} \leq |r_{ij} + r_{jk}|\} \quad (1)$$

While the configuration of the system is uniquely defined by these distances, not all distances are allowable. This means that there are geometric constraints between some of the distances independent of the finite density assumption. Thus it is sometimes easier to define a unique configuration where the restriction is clearly obvious. One such is to specify the distances between two of the bodies and the angle between these two bodies centered on the third body (see Fig. 6). Thus, denoting the bodies with the unique indexing*i*, *j*, *k*, the configuration can be specified as

$$q = \{r_{ij}, r_{jk}, \theta_{ki} \mid r_{lm} \geq (R_l + R_m)\} \quad (2)$$

where the final distance rkirki can be explicitly computed from the cosine rule:

$$r_{ki}^2 = r_{ij}^2 + r_{jk}^2 - 2 r_{ij} r_{jk} \cos\theta_{ki} \quad (3)$$

Note that the angle θki will also have constraints placed upon it, as the associated length must satisfy rki≥Ri+Rk. These two expressions of the third degree of freedom, θkiθki or rki, will be used equivalently.

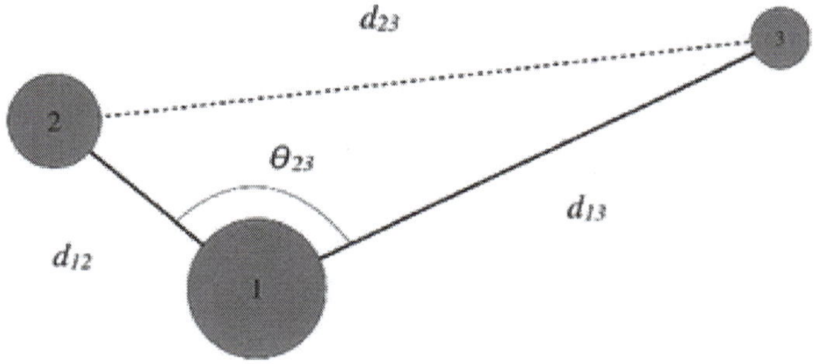

Figure 6. Configuration of the system

There are additional degrees of freedom of the triad of bodies with respect to inertial space, which can be tracked by the unit vector of the angular momentum, ***H***, which are briefly discussed later.

The gravitational potential is equivalent to the three-body point-mass potential due to the symmetry of spherical mass distributions.

$$\mathcal{U} = -\mathcal{G}\left[\frac{M_1M_2}{r_{12}} + \frac{M_2M_3}{r_{23}} + \frac{M_3M_1}{r_{31}}\right] \quad (4)$$

The moment of rotational inertia of each sphere is equal to $2M_iR_i^2/5$. For a given placement of the three masses, the total inertia dyad of the system can be constructed as

$$\boldsymbol{I} = \frac{1}{M_1+M_2+M_3}\sum_{i=1}^{2}\sum_{j=2}^{3} M_iM_j\left[r_{ij}^2\boldsymbol{U} - \boldsymbol{r}_{ij}\boldsymbol{r}_{ij}\right] + \sum_{i=1}^{3}\frac{2M_iR_i^2}{5}\boldsymbol{U} \quad (5)$$

where ***U*** is the identity dyad. Note that this form uses the Lagrange identity and assumes that the center of mass is nominally at a zero point. The inertia matrix is orientable, but for the amended potential only its orientation relative to the constant angular momentum vector direction, **H^**, is needed. Dotting the dyad on both sides by this unit vector yields

$$I_H = \frac{1}{M_1+M_2+M_3}\sum_{i=1}^{2}\sum_{j=2}^{3} M_iM_j\left[r_{ij}^2 - (\hat{\boldsymbol{H}}\cdot\boldsymbol{r}_{ij})^2\right] + \sum_{i=1}^{3}\frac{2M_iR_i^2}{5} \quad (6)$$

The principal moments of inertia for a three point-mass system have the following relation: $I_{max} = I_{int} + I_{min}$. Furthermore the maximum moment of inertia of the point masses will always be perpendicular to the plane that the three bodies form (Greenwood 1988). Thus, with our assumption that the body spins about its maximum moment of inertia, the quantities $(\hat{H} \cdot r_{ij}) = 0$ and the moment of inertia simplifies to

$$I_H = M_1 M_2 r_{12}^2 + M_2 M_3 r_{23}^2 + M_3 M_1 r_{31}^2 + I_S \quad (7)$$

$$I_S = \frac{2}{5} M_1 R_1^2 + \frac{2}{5} M_2 R_2^2 + \frac{2}{5} M_3 R_3^2 \quad (8)$$

For rotation about the intermediate and minimum moments of inertia, we note that IHIH will be strictly less than or equal to this above value, with equality between the intermediate and maximum only occurring when the minimum moment of inertia of the three particles (without the rigid sphere contributions) is 0. The maximum moment of inertia of the point masses can never be zero, due to the finite size of the particles.

3.3 Normalization

To simplify the discussion, normalize the system with a length and a mass scale. The length scale used $R_T = R_1 + R_2 + R_3$ while the mass scale is $M_T = M_1 + M_2 + M_3$. Denote $m_i = M_i/M_T$, $r_i = R_i/R_T$, and $d_{ij} = r_{ij}/R_T$. In normalized coordinates the fundamental quantities take on the values

$$\mathcal{U} = -\left[\frac{m_1 m_2}{d_{12}} + \frac{m_2 m_3}{d_{23}} + \frac{m_3 m_1}{d_{31}}\right] \quad (9)$$

$$I_H = m_1 m_2 d_{12}^2 + m_2 m_3 d_{23}^2 + m_3 m_1 d_{31}^2 + I_S \quad (10)$$

$$I_S = \frac{2}{5} m_1 r_1^2 + \frac{2}{5} m_2 r_2^2 + \frac{2}{5} m_3 r_3^2 \quad (11)$$

with the angular momentum being normalized by the dividing factor $\sqrt{\mathcal{G}M_T^3 R_T}$ and the energy normalized by the dividing factor $\mathcal{G}M_T^2/R_T$. For both *H* and *E* the same notational designation is kept for the normalized values.

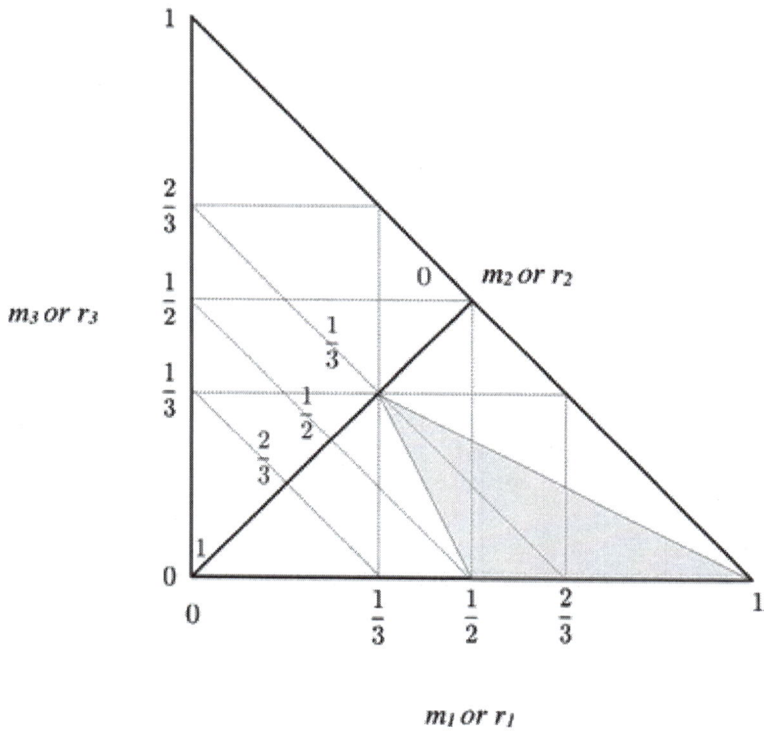

Figure 7. Triangle defined for the bodies with the region of study *shaded*

The normalizations provide two identities:

$$r_1 + r_2 + r_3 = 1 \tag{12}$$

$$m_1 + m_2 + m_3 = 1 \tag{13}$$

There are also fundamental relationship between the riri and the mimi, assuming constant density.

$$m_i = \frac{r_i^3}{r_1^3 + r_2^3 + r_3^3} \quad (14)$$

$$r_i = \frac{m_i^{1/3}}{m_1^{1/3} + m_2^{1/3} + m_3^{1/3}} \quad (15)$$

3.4 Parameterization of the Problem

Any given variant of the F3BP can be identified with a point in a compact, two-dimensional triangle, using either the masses or the radii. Plotting the radii *r1,r3* or masses *m1,m3* along two perpendicular axes each of them can only take values between 0 and 1, and that furthermore they will be bounded by a diagonal defined by $r_1 + r_3 \leq 1$ or $m_1 + m_3 \leq 1$. On the boundary of this equality r2=m2=0. If, instead, a diagonal defined by r1+r3=r13<1 or $m_1 + m_3 = m_{13} < 1$ is drawn, then the value of the second radius or mass will $r_2 = 1 - r_{13}$ or $m_2 = 1 - m_{13}$. This also lends itself to a graphical description, shown in Fig. 7 for the masses. Every point within this triangle defines a unique F3BP in terms of its relative masses. This paper studies the bifurcation structure across the entire region, however, due to the symmetry of the problem the study can be restricted to a specific region only. To that end, consider the restrictions

$$0 \leq m_3 \leq m_2 \leq m_1 \leq 1 \quad (16)$$

$$0 \leq r_3 \leq r_2 \leq r_1 \leq 1 \quad (17)$$

This region is shaded in Fig. 7. There are five other equivalent triangles defined by reordering the different inequalities given above. The approach taken will be to exhaustively study all possible relative equilibria in the denoted region, the results of which can then be easily applied to all other regions.

With this convention, there are additional constraints for the masses and radii.

$$\frac{1}{3} \leq (r_1, m_1) \leq 1 \tag{18}$$

$$0 \leq (r_3, m_3) \leq \frac{1}{3} \tag{19}$$

$$0 \leq (r_3, m_3) \leq (r_2, m_2) \leq \frac{1}{2} \tag{20}$$

Previous research has exhaustively explored the bifurcation structure and properties for two general cases along the boundary of this triangle. One is at the point (1/3, 1/3, 1/3), when all masses and sizes are equal (Scheeres 2012). In this case a more limited number of relative equilibria were found with a less complex bifurcation structure. The other case is for m3=0, in essence just considering the two mass case with $0 \leq m_2 \leq 1/2 \leq m_1 \leq 1,$ along the base of the triangle (Scheeres 2012). In this region the number of relative equilibria are also much fewer and the bifurcation structure less complex.

For the problem we study, the spherical three-body problem, we can easily just consider the planar motion of the system, with rotation occurring about a principal moment of inertia of the system. We note that the spheres contribute to the system's total angular momentum but have the same moment of inertia about any axis. In general we will assume that the system rotates about the maximum moment of inertia, but will justify this later.

4 BACKGROUND AND SUPPORTING RESULTS

A few definitions and supporting Lemmas are stated for use in this paper. Some of these are classical results while others have been considered more recently Scheeres (2012, in press), and thus the proofs are only briefly reviewed to point out their salient features. Specific results for our current analysis are worked out in detail.

Lemma 1

The Total Energy of the system is conserved in the absence internal dissipation and equals E=Tr+E, where *E* is the total energy and Tr is kinetic energy of the system components relative to each other, evaluated in the rotating frame with inertial angular velocity ***H*/ *IH***.

Proof

For rotation about a principal axis of the system, EE equals the amended potential as introduced by Smale (1970a, (1970b), and specifically considered by Arnold for the three-body problem in Arnold et al. (1988), pp. 66–67. For a system rotating about its principal axis the proofs in Scheeres (in press) apply, showing that the amended potential arises from a Routh reduction of the system. The Routhian is shown to have a Jacobi integral, which is identical to the total energy of the system.

Lemma 2

The total energy of the system is strictly bounded from below by the amended potential: $\mathcal{E} \leq E$. *If* $E = \mathcal{E}$, *then* $T_r = 0$. If the system is momentarily stationary (Tr=0) and spins about a principal axis of inertia of the system, then E=E. Thus the inequality is sharp and the lower bound can be achieved.

Proof

Lemma 2 is proven in Scheeres (in press). The proof establishes the inequality using the Cauchy Inequality applied to the angular momentum, and shows that it is sharp through direct construction.

Another important feature of the system involves the existence of minimum energy configurations for full-body systems. The following lemma establishes the existence of minimum energy states for all values of angular momentum. This result also provides the fundamental motivation for the current study.

Lemma 3

For a finite density distribution, the amended potential EE has a global minimum for all values of angular momentum *H*.

Proof

The proof is given in Scheeres (2012) and involves showing that E is compact and bounded over all possible values of the configuration space. This requires the finite density assumption, as this blocks individual point masses from coming arbitrarily close to each other. If body *i* escapes to ∞∞∞ relative to bodies *j* and *k*, then the amended potential takes on the value E=UjkE=Ujk and remains bounded in the interval $[-m_j m_k/(1-r_i), 0]$. If all three bodies escape to ∞∞∞ relative to each other, then E=0.

Given the definition of the amended potential and its properties relative to the total energy of the system, the relative equilibrium and energetic stability can be defined. Following this conditions under which these are satisfied are stated.

Definition 2

Relative Equilibrium A given configuration q∗ is said to be a "Relative Equilibrium" if its internal kinetic energy is null (Tr=0), meaning that E=E at an instant, and if it remains in this state over at least a finite interval of time.

Definition 3

Energetic Stability A given relative equilibrium qq∗ is said to be "Energetically Stable" if any equi-energy deviation from that relative equilibrium requires a negative internal kinetic energy, Tr<0, meaning that this motion is not allowed.

Lemma 4

Consider a system with an amended potential EE as defined above with *n* degrees of freedom, *m* of which are activated in such a way that only the variations $\delta q_j \geq 0,\ j = 1, 2, \ldots, m$ are allowed. The degrees of freedom qi for $m < i \leq n$ *are free.*

The necessary and sufficient conditions for a system in a configuration qq∗ to be in a relative equilibrium are that at this configuration:

1. $T_r = 0$

2. $\mathcal{E}_{q_j} \geq 0 \; \forall \; 1 \leq j \leq m$

3. $\mathcal{E}_{q_i} = 0 \; \forall \; m < i \leq n$

The necessary and sufficient conditions for a system in a relative equilibrium to be energetically stable are that:

1. $\mathcal{E}_{q_j} > 0 \; \forall \; 1 \leq j \leq m$

2. $\left[\frac{\partial^2 \mathcal{E}}{\partial q_i \partial q_k}\right] > 0 \; \forall \; m < i, k \leq n$

Proof

The proof of Lemma 4 is found in Scheeres (in press). It relies on taking variations of the amended potential, asserting the principle of conservation of energy and using the general form of the Lagrange equations of motion for the full-body system.

Next a few results of relevance for the system are stated regarding bifurcation of relative equilibria and their stability. Specific example bifurcations and their properties are shown in Fig. 8

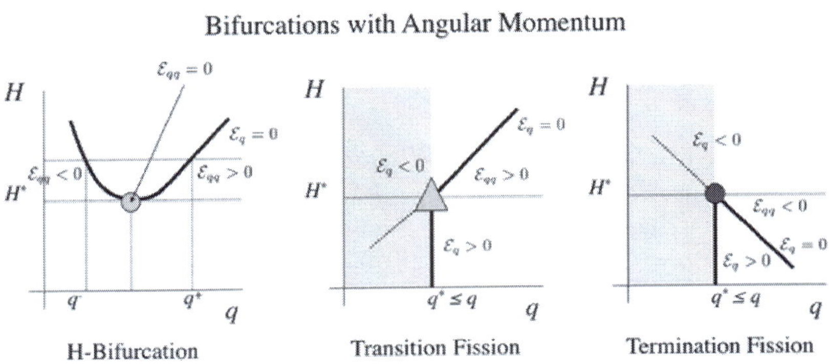

Figure. 8. Examples of bifurcations of interest in this problem, and their stability properties

Definition 4

Symmetric Bifurcation A bifurcation of two relative equilibria which follow a symmetric path relative to each other about a reflection line at changing values of angular momentum.

Definition 5

H-Bifurcation An H-Bifurcation occurs when, under increasing angular momentum, a pair of relative equilibria appear in a degree of freedom q that is not at a constraint. At its first appearance there must be a degeneracy of the form Eqq=0 that will generically disappear under increasing angular momentum.

Definition 6

Fission A collection of bodies in a relative equilibria with at least one active constraint is said to "fission" if, under an increase in angular momentum, the active constraint is released, meaning that a free relative equilibria intersects with it. Following fission the body either may transition into a new relative equilibrium without that active constraint or may no longer lie in any relative equilibrium associated with that configuration.

Definition 7

Termination Fission A fission bifurcation where the relative equilibria disappear at higher values of angular momentum.

Definition 8

Transition Fission A fission bifurcation where the relative equilibria continues with its constraint inactive at higher values of angular momentum.

Now a particularly useful lemma is proven, which enables us to relate the stability of equilibrium points to how their coordinate changes as a function of angular momentum H.

Lemma 5

Assume a relative equilibria exists with a 1–1 relationship between a single degree of freedom q and the angular momentum H, meaning that along the local family of relative equilibria as the angular momentum is changed only the degree of freedom q changes.

Then sign$(\mathcal{E}_{qq}) = \text{sign}(\partial H/\partial q)$ at the relative equilibria. Thus, if $\partial H/\partial q < 0$ the equilibrium point will be energetically unstable and if $\partial H/\partial q > 0$ it could be stable, depending on the other degrees of freedom.

Proof

From the lemma statement it can be assumed that all other degrees of freedom lie in a relative equilibrium condition independent of the local value of *H*. Given this, assume that there exists a value q∗q∗ such that the scalar equation $\mathcal{E}_q|_* = -H^2/(2I_H^2)I_{Hq} + \mathcal{U}_q = 0.$ This can be solved for $H^* = \left(I_H\sqrt{2I_{Hq}/\mathcal{U}_q}\right)\Big|_*,$ where the righthand side is a function of q∗ and by assumption is non-singular. Now consider a neighboring relative equilibrium at a different value of *H* and hence *q*, with the values defined locally by the expansion

$$\mathcal{E}_q(H^* + \Delta H, q^* + \Delta q) \;=\; 0 + \mathcal{E}_{qH}|_* \left.\frac{\partial H}{\partial q}\right|_* \Delta q + \mathcal{E}_{qq}|_* \Delta q + \ldots$$

Setting this to zero and solving for an arbitrary Δq yields

$$\mathcal{E}_{qq}|_* \;=\; -\,\mathcal{E}_{qH}|_* \left.\frac{\partial H}{\partial q}\right|_*$$

However, from the defining equation for EqEq given above, it is seen that $\mathcal{E}_{qH} = -H/I_H^2 I_{Hq},$ where IHq>0IHq>0by inspection of Eq. (10). Thus, the sign of Eqq|∗Eqq|∗ equals the sign of $\left.\frac{\partial H}{\partial q}\right|_*$ □⍰

Corollary 1

For an increasing angular momentum *H*, a free relative equilibria that ends in a Termination Fission is always unstable in the degree of freedom *q*. Conversely, a free relative equilibria that emanates from a Transition Fission is always stable in the degree of freedom *q*.

Proof

Assume the active constraint is defined to be q=0. By definition, a Termination Fission occurs when a relative equilibrium at $q^* > 0$ moves toward the general constraint q=0 under increasing angular momentum. Thus $\partial H/\partial q < 0$ and from Lemma 5 the relative equilibrium is unstable. Conversely, a Transition Fission occurs when a relative equilibrium at q∗≥0 moves away from the general constraint q=0 under increasing angular momentum. Thus $\partial H/\partial q < 0$ and from Lemma 5 the relative equilibrium is energetically stable in the degree of freedom *q*, although it may be unstable in other degrees of freedom.

Finally, we end with a Lemma on the rotation axis that a stable configuration must have.

Lemma 6

Any relative equilibrium configuration not rotating about the maximum moment of inertia of the body will be energetically unstable.

Proof

If a body is in a relative equilibria it must rotate about a principal moment of inertia. It can then be treated as a rigid body, at least up to first-order variations in its internal configuration and inertial orientation. If it is not rotating about its maximum moment of inertia, it must be rotating about its intermediate or minimum moment of inertia. The relevant total energy of the function system then equals $H^2/(2I_i)$ where IiIi is a principal moment of inertia (here ignoring internal variations). From the classical Poinsot construction the body will be at a saddle point of the energy function if rotating about the intermediate axis and will be at a local maximum of the energy function for rotating about the minimum axis. In either case, the rigid body rotation is not stable in the energetic sense as it can depart from this rotation axis while conserving energy with an increase in kinetic energy.

Due to this result, we only consider rotation about the maximum moment of inertia, which will always lie in the plane containing the three bodies. In the degenerate case where the bodies are in a line, the system will rotate perpendicular to its line of symmetry.

With these Definitions, Lemmas and Corollaries stated, the relative equilibria and stability of the full three-body problem can be established.

5 Existence, Stability and Bifurcation of Relative Equilibria

In this section, having stated the theorem and developed the necessary background, the detailed proof of Theorem 1 is now given.

Proof

To systematically explore the existence, stability and bifurcation of the relative equilibria the systems with different numbers of degree of freedom constraints activated and conditions for these to be released are considered separately. The discussion starts with all three DOF constraints activated and progressing to fewer and fewer until all degrees of freedom are not constrained. The "Appendix" contains the detailed partial derivatives and variation conditions of the amended potential needed for the following discussions.

5.1 Three Active Constraints: Lagrange Resting Configurations

Existence For the three constraints to be active requires that $d_{ij} = r_i + \quad r_j = 1 - r_k$ for all of the indices. This configuration can only occur when the three bodies are mutually resting on each other. The relative angle between adjacent grains are then defined by

$$\cos\theta_{ki} = \frac{(1-r_k)^2 + (1-r_i)^2 - (1-r_j)^2}{2(1-r_k)(1-r_i)} \tag{21}$$

$$\sin\theta_{ki} = \frac{\sqrt{(1-r_k)^2(r_k - r_i r_j) + (1-r_i)^2(r_i - r_j r_k) + (1-r_j)^2(r_j - r_k r_i)}}{\sqrt{2}(1-r_k)(1-r_i)} \tag{22}$$

(22)

where *k*, *i* take on all possible values. The corresponding values of IHIH and UU in this configuration are

$$I_H = m_i m_j (1-r_k)^2 + m_j m_k (1-r_i)^2 + m_k m_i (1-r_j)^2 + I_S$$

(23)

$$\mathcal{U} \quad = \quad -\left[\frac{m_i m_j}{1-r_k} + \frac{m_j m_k}{1-r_i} + \frac{m_k m_i}{1-r_j}\right] \tag{24}$$

There are two unique orderings of the resting configuration, mirroring the orbital Lagrange configuration, which results in two distinct relative equilibria. Due to this these configurations are called the Lagrange Resting (LR) configurations.

Stability As this is the minimum distance for each of these bodies to achieve, this also implies that the potential energy will be minimized at this configuration. From this it can immediately be concluded that for H=0 this particular resting configuration is the minimum energy configuration of the system and hence is stable.

Bifurcation As *H* increases from zero this system should exist as a relative equilibrium for some range of *H*, but to discover the precise range when this holds requires that the transition from three to two active constraints be investigated. Thus, as angular momentum is increased, conditions for when one of these constraints is no longer enforced is sought, meaning that one of the degrees of freedom will have an allowable variation that decreases the energy. For this configuration each of the three distances can be tested in turn to see which will lose positivity first. For the condition tested, consider the angle variation δθki≥0 keeping the other two constraints δdij=δdjk=0. The condition for existence (and stability) of this configuration then becomes δθkiE≥0. Evaluating this explicitly and substituting for the equal resting conditions yields

$$\delta_{\theta_{ki}}\mathcal{E} \quad = \quad m_k m_i\left[-\frac{H^2}{I_H^2} + \frac{1}{(1-r_j)^3}\right](1-r_k)(1-r_i)\sin\theta_{ki}\delta\theta_{ki} \tag{25}$$

and substituting in for sinθkisin□θki yields

$$\delta_{\theta_{ki}}\mathcal{E} \quad = \quad \frac{m_k m_i}{\sqrt{2}}\left[-\frac{H^2}{I_H^2} + \frac{1}{(1-r_j)^3}\right] \times \tag{26}$$
$$\sqrt{(1-r_k)^2(r_k - r_i r_j) + (1-r_i)^2(r_i - r_j r_k) + (1-r_j)^2(r_j - r_k r_i)\delta\theta_{ki}}$$

(26)

Changing *k*, *i* for *i*, *j* and *j*, *k* only changes the items on the first line, and thus the controlling condition for the existence and stability of the Lagrange Resting configurations is

$$\frac{1}{(1-r_j)^3} > \frac{H^2}{I_H^2} \tag{27}$$

which must hold for j=1,2,3j=1,2,3. Thus the minimum value of rjrj gives the minimum value of *H* for the inequality to be violated. For the specified definitions this means that j=3j=3 and the loss of stability occurs about the angle θ12θ12, meaning that the Lagrange Resting configuration will undergo a Termination Fission by losing contact between its two largest bodies, pivoting about the smallest grain (see Fig. 3).

5.2 Two Active Constraints

In this case two bodies rest on each other, but do not have the third contact active. A convenient way to express this is to have the two distances at their minimum value and leave the angle free, or $d_{ij} = 1 - r_k,\ d_{jk} = 1 - r_i$ with θ_{ki} only constrained by the resting limit, dki≥1−rj. For the moment assume that Edjk>0 (this will be checked later), and thus there is only one degree of freedom to be concerned with. Taking the first variation and substituting for the distances yields

$$\delta_{\theta_{ki}}\mathcal{E} = m_k m_i \left[-\frac{H^2}{I_H^2} + \frac{1}{d_{ki}^3}\right](1-r_k)(1-r_i)\sin\theta_{ki}\delta\theta_{ki} \tag{28}$$

which must now be identically equal to zero for the system to be in equilibrium. There are two possibilities, sin $\theta_{ki} = 0$ or $-\frac{H^2}{I_H^2} + \frac{1}{d_{ki}^3} = 0.$ Both can occur and are discussed separately, the former is called the Euler Resting configuration and the latter the Transitional Resting configuration. No assumptions are made about the ordering of the bodies in terms of mass, unless specified.

Each case must be tested for when the configurations cease to exist, which will occur once one of the energy variations in the active distance constraints equals zero. These will be explicitly tested for each case to determine conditions at which these equilibria no longer exist.

5.2.1 Euler Resting Configurations

Existence First consider the case when θki=π, noting that the angle cannot equal zero due to the finite radius constraints. Then the first variation is identically equal to zero and the bodies rest on a straight line with the ordering *i, j, k*, the system forming a relative equilibrium. These are notationally denoted as ER*ijk*, noting that configuration ER*kji* is considered to be equivalent. Now the moment of inertia and potential energy take on the values

$$I_H = m_i m_j (1 - r_k)^2 + m_j m_k (1 - r_i)^2 + m_k m_i (1 + r_j)^2 + I_S \tag{29}$$

$$\mathcal{U} = -\left[\frac{m_i m_j}{1 - r_k} + \frac{m_j m_k}{1 - r_i} + \frac{m_k m_i}{1 + r_j}\right] \tag{30}$$

with the main difference from the Lagrange Resting (LR) configurations being that the distance dki=1+rjdki=1+rjdue to the elongate geometry.

Stability Under the assumption that the two distance variations are both positive (which is true for a low enough value of *H*), the stability of this relative equilibrium can be analyzed by computing the second-order variation evaluated at the resting configuration.

$$\delta_{\theta_{ki}\theta_{ki}} \mathcal{E} = -m_k m_i \left[-\frac{H^2}{I_H^2} + \frac{1}{(1 + r_j)^3}\right] (1 - r_k)(1 - r_i) \delta\theta_{ki}^2 \tag{31}$$

Stability of this configuration occurs when $\delta_{\theta_{ki}\theta_{ki}} \mathcal{E} > 0$ which places a lower limit on the angular momentum for stability

$$\frac{I_H^2}{(1 + r_j)^3} < H^2 \tag{32}$$

Note that the value of angular momentum is lower than the angular momentum at which the LR configurations cease to exist. Also, the stability transition occurs when the Transitional Resting configuration conditions are satisfied for the same configuration, indicating that a bifurcation occurs.

Bifurcation For lower values of angular momentum the Euler Resting configuration exists, but is unstable and mimics an inverted pendulum. When the stability condition is satisfied, the system mimics a hanging pendulum and will remain stable until one of the energy distance variations becomes zero, indicating a transition from two active constraints to a single active constraint. To probe when this occurs, substitute the equilibrium condition into Eq. (86) to find

$$\begin{aligned}\delta_{d_{ij}}\mathcal{E} &= m_i\left\{m_j\left[-\frac{H^2}{I_H^2}+\frac{1}{(1-r_k)^3}\right](1-r_k)\right.\\ &\left.+m_k\left[-\frac{H^2}{I_H^2}+\frac{1}{(1+r_j)^3}\right](1+r_j)\right\}\delta d_{ij}\end{aligned} \tag{33}$$

Setting this to be greater than or equal to zero defines when the ER relative equilibrium configuration exists, and can be solved for as a condition on angular momentum

$$[m_j(1-r_k)+m_k(1+r_j)]\frac{H^2}{I_H^2} \le m_j\frac{1}{(1-r_k)^2}+m_k\frac{1}{(1+r_j)^2} \tag{34}$$

The precise value of *H* when this is first violated is discussed in a later section. For the current analysis it suffices to note that this inequality is always satisfied when the ER configurations first become stable. Substituting (H/IH)2=1/(1+rj)3 and simplifying yields

$$0 \le r_j + r_k \tag{35}$$

which is trivially satisfied for any *j* or *k*. It is also clear that a large enough *H* will always be able to violate the existence condition. Generically, one of the two bodies *i* or *k* will separate from *j*, leaving the other body in contact and transitioning the configuration into the Euler Aligned configuration.

5.2.2 Transitional Resting Configurations

Existence When the Euler Resting (ER) configurations becomes stable, a pair of solutions that satisfy the second equilibrium condition bifurcate from or into the resting configuration. The condition in general is H2/dki3, but now the moment of inertia ***IH*** becomes a function of θki and must change with *H* to maintain this condition. There are two branches, $\theta_{ki} > \pi$ and $\theta_{ki} < \pi$, and these give two different orderings of the configuration—ultimately corresponding to the two different orientations of several of the equilibrium configurations. The moment of inertia and potential energy now take on the more generalized form

$$I_H = m_i m_j (1 - r_k)^2 + m_j m_k (1 - r_i)^2 + m_k m_i d_{ki}^2 + I_S \tag{36}$$

$$\mathcal{U} = -\left[\frac{m_i m_j}{1 - r_k} + \frac{m_j m_k}{1 - r_i} + \frac{m_k m_i}{d_{ki}}\right] \tag{37}$$

$$d_{ki}^2 = (1 - r_k)^2 + (1 - r_i)^2 - 2(1 - r_k)(1 - r_i)\cos\theta_{ki} \tag{38}$$

Stability Evaluating the second variation of the energy with respect to θki yields

$$\delta_{\theta_{ki}}^2 \mathcal{E} = m_k m_i \left[4 m_i m_k d_{ki}^2 - 3 I_H\right] \frac{(d_{ij} d_{jk} \sin\theta_{ki})^2}{I_H d_{ki}^5} (\delta\theta_{ki})^2 \tag{39}$$

Stability, when the configuration exists, then hinges on the sign of 4mimkdki2−3IH. Making the substitution from Eq. (14) the stability condition can be reduced to

$$d_{ki}^2 > \frac{3}{r_i^3 r_k^3}\left[\frac{2}{5}\left(r_i^3 + r_j^3 + r_k^3\right)\left(r_i^5 + r_j^5 + r_k^5\right) + r_i^3 r_j^3 (1 - r_k)^2 + r_j^3 r_k^3 (1 - r_i)^2\right] \tag{40}$$

where 1−rj≤dki

Note that the equilibrium configuration does not necessarily exist across this entire range of mutual distances. Specifically, the distance variation conditions must be verified for the configuration to exist. Substituting the equilibrium condition into Eq. (86) then yields the existence condition (after simplification)

$$\frac{H^2}{I_H^2} \leq \frac{1}{(1-r_k)^3} \tag{41}$$

where *k* is the radius of either of the outer resting bodies. Note that the transitional resting configurations will always exist over some interval of angular momentum, as substituting the initial bifurcation conditions of (H/IH)2=1/(1+rj)3 can be trivially shown to satisfy the above existence condition. Again, note that*H* can also always be chosen large enough for the existence condition to be violated. There are three different possible situations to cover, investigated in detail below.

$i = 1, j = 2, k = 3$ For this sequence the Transitional Resting equilibrium are unstable and migrate from the ER123 configuration (which they stabilize upon bifurcation from it) to the distance d31=1-r3, at which point body 1 separates from the system. To determine instability evaluate Eq. (40) over the entire range of radius values and verify that it is never satisfied. To see this consider the contact conditions from Eqs. (86) and (87). For these conditions to hold both must be greater than or equal to zero for a positive variation in the mutual distance dij and djk. Substituting the equilibrium condition (H/IH)2=1/d331 and simplifying, the condition for existence of the TR123 configuration is that both

$$d_{31} \geq d_{12} \tag{42}$$

$$d_{31} \geq d_{23} \tag{43}$$

For the current configuration, $d_{12} = 1 - r_3 > d_{23} = 1 - r_1$. Thus the controlling condition is $d_{31} \geq 1 - r_3$. Now note that $d_{31} \geq 1 - r_2$ and that $1 - r_3 > 1 - r_2$, thus this inequality is violated prior to the TR123 configuration reaching the LR configuration, and as noted occurs once $d_{31} = 1 - r_3$. See Fig. 2 for the evolutionary path for this situation.

$i = 1, j = 2, k = 3$ For this sequence the Transitional Resting equilibrium are unstable (determined as before) and migrate from the ER312 configuration (which they stabilize upon bifurcation from it) to the distance d23=1−r3 when body 2 separates from the system. Similar to above, the condition for existence of the TR312 configuration is that

$$d_{23} \geq d_{31} \tag{44}$$

$$d_{23} \geq d_{12} \tag{45}$$

For the current configuration, d31=1−r2. Thus the controlling condition is d23≥1−r3. Now note that d23≥1−r1 and that 1−r3>1−r1, thus this inequality is violated prior to the TR312 configuration reaching the LR configuration, and as noted occurs once d23=1−r3 again. See Fig. 4 for the evolutionary path for this situation.

$i = 1, j = 2, k = 3$ For this sequence the Transitional Resting equilibria exist across the range of radius limits, going from ER132 to Lagrange Resting configurations. For this configuration there are ranges of parameters for which there are stable relative equilibria. To identify these regions compare the upper inequality limit to when the distance for stability is less than the maximum distance 1+rj. Plotting out this region delineates the small oval region in Fig. 9. For parameter values within this region the evolution of the TR132 configuration becomes more complex. Specifically, the angular momentum profile in this region is such that there are two relative equilibria defined at a given level of angular momentum, one toward the LR configuration (which is always at a local maximum of the energy and thus is unstable) and one toward the ER132 configuration (which becomes a local minimum of the energy and thus is stable). Figure 3 shows the two different pathways that can occur.

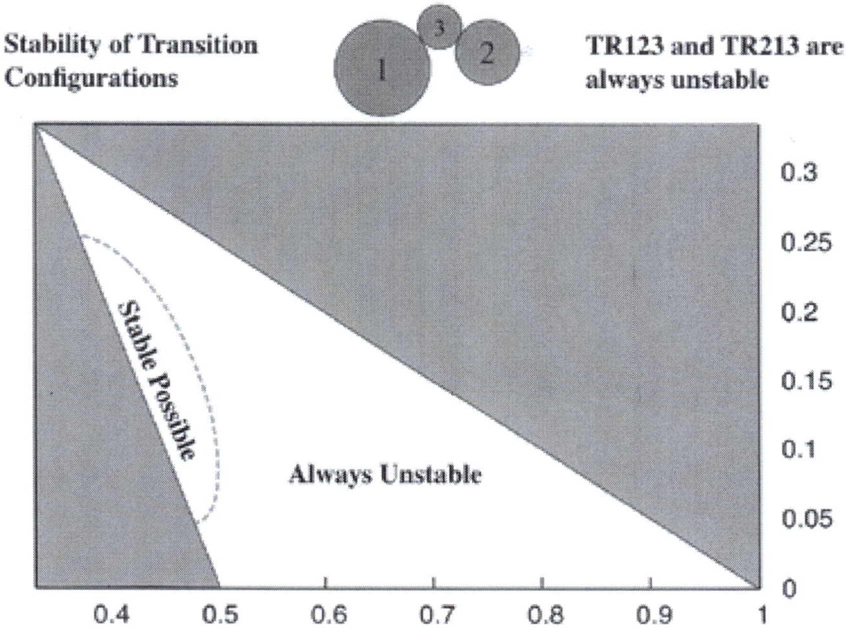

Figure 9. Region where stable TR132 configurations can lie. The horizontal axis is the radius r1r1 and the vertical axis is the radius r3r3. The same axes are used for all subsequent figures

Bifurcation For the TR configurations which are always unstable, as the angular momentum is increased they first bifurcate into existence by stabilizing the ER configurations. Then as H is increased they migrate toward more compact configurations. The TR123 and TR312 configurations then end with one of the bodies separating from the other two. The TR132 configuration migrates all the way to the LR configuration and destabilizes it, thus terminating both the LR and TR132 configurations.

For TR configurations that can be stable, indicated in Fig. 9, the sequence is different. Here as H is increased an H-Bifurcation occurs at an angle between the minimum (or maximum) constrained value of θ12 and π. To show this consider the angular momentum as a function of distance d12, H=IH(d12)/d3/212 Taking the partial of this with respect to d12 shows that there is a zero in the interval 12≤1+r3 (meaning that H takes on an extreme value) whenever the stability condition in Eq. (40) is satisfied. Further, taking the second partial of H and substituting the equilibrium condition shows that this is always positive, meaning that *H* takes on a minimum value in the interval. Thus, as *H* is increased the two equilibria exist on either side of the minimum, with no other equilibria emerging due to the definiteness of *H* as a function of d12. From Lemma 5 the equilibria that moves down to the LR configuration must be unstable, and thus the equilibria that moves toward the ER configuration is stable (this also agrees with the condition as formulated in Eq. (40)).

Thus the unstable TR132 configuration continues down to the LR configuration and terminates it. The stable TR132 configuration moves toward the ER132 configuration and terminates there, stabilizing the ER132 configuration. The existence of these stable TR132 configurations is unexpected and breaks the symmetry otherwise seen in these configurations. The region where these occur correspond to grains with a nearly equal r1 and r2, with r3 neither close to zero or to the size of the other grains.

5.3 One Active Constraint

Now consider relative equilibria when there is a single active constraint. In this configuration two of the bodies rest on each other, say *i* and *j* and thus dij=1−rk, and the third body is located by the distance djkdjk and by either dkidki or the angle θki There are two classes of relative equilibrium solutions in this class, with the two bodies in contact either being aligned with the third body, or with their line of contact being orthogonal to the third body. The former are called the Euler Aligned configurations and the latter the transverse, or Isoceles, configurations—the terminology arising due to the structure that these make. For these structures there are two limiting cases, one where the final active constraint separates and the other where one of the free constraints becomes activated. The former occurs when the single active constraint configurations intersect with the orbital configurations. The latter occurs when it intersects with a double-active configuration. These two classes of configurations are discussed in turn.

First it can be established that these are the only relative equilibrium configurations. Consider Eqs. (87) and (88) in the "Appendix," which both must equal zero. There are two possibilities for Eq. (88) to equal zero, $\sin\theta_{ki} = 0 \text{ or } H = I_H/d_{ki}^{3/2}$. Consider $\theta_{ki} = \pi,$ as setting the angle to 0 is equivalent to a reordering of the bodies. Then for Eq. (87) to equal zero the condition becomes $m_j \left(H^2/I_H^2 - 1/d_{jk}^3\right) d_{jk} + m_i \left(H^2/I_H^2 - 1/d_{ki}^3\right) d_{ki} = 0.$ In this configuration dki>djk and thus along this configuration it can never occur that $H^2/I_H^2 = 1/d_{ki}^3,$ meaning that this condition will not intersect with the θki=π configuration. If *H* is chosen such that IH/dki3/2 it is possible for the second condition to be satisfied, which is explored in more detail below.

The alternate condition to consider is $H^2/I_H^2 = 1/d_{ki}^3,$ with no immediate constraint on θki Then, by substitution into Eq. (87) yields the condition H2/IH2=1/djk3, or ki=djk=d. From Eq. (3) and with dij=1−rk the condition on θki becomes

$$\cos\theta_{ki} = \frac{1}{2}\frac{1-r_k}{d} \tag{46}$$

Note that *d* can always be chosen large enough for θki to be well defined.

5.3.1 Isosceles Configurations

Existence The isosceles configurations are described by having two grains in contact, nominally *i* and *j*, the third grain *k* in non-contact with these grains, and with the line connecting the grains *i* and *j* being perpendicular to the line from grain *k* to the center of mass of grains *i* and *j*. These are referred to as IS*ij*-*k*, with the first two indices indicating the grains in contact and the separated third index the separated grain. In terms of Eqs. (86, 87) and (88), set $d_{ij} = 1 - r_k \text{ and } d_{jk} = d_{ki} = d,$ forming an Isosceles triangle. The equilibrium condition is then simply stated as

$$\frac{H^2}{I_H^2} = \frac{1}{d^3} \tag{47}$$

$$I_H = m_i m_j (1-r_k)^2 + (m_j m_k + m_k m_i) d^2 + I_S \tag{48}$$

Making this substitution, see that $\delta_{d_{jk}}\mathcal{E} = \delta_{\theta_{ki}}\mathcal{E} = 0.$. So long as djk≥1−ri and dki≥1−rj, the remaining condition for this equilibrium to be satisfied is δdijE≥ which can be simplified to the condition

$$d = d_{ki} = d_{kj} \geq 1 - r_k \tag{49}$$

Now consider the existence of each of the possible combinations, in turn.

IS12-3 Here the grains in contact are separated by a distance 1−r3 and the controlling distance of the equal legs of the triangle will be d31=1−r2. Note that 1−r2, and thus the above existence condition will be violated when d31=d23=1−r3, and in fact the three grains will lie at the vertices of an equilateral triangle. This condition corresponds to the intersection of IS12-3 with the orbital Lagrange configuration, LO, and terminates the IS12-3 configuration, without having grain 3 contacting the other two grains. This sequence is isolated from the previous configurations as the three grains never come into contact and is shown in Fig. 5.

IS23-1 Now the grains in contact are separated by a distance 1−r1 and the controlling distance of the equal legs of the triangle will be d12=1−r3. Now as 1−r3>1−r1, grains 1 and 2 will touch prior to the grains 2 and 3 separating. Once grains 1 and 2 touch the configuration matches the end-state configuration of the TR123 configuration. Thus the TR123 and IS23-1 configurations terminates, as shown in Fig. 2.

IS31-2 Now the grains in contact are separated by a distance $1 - r_3$ and the controlling distance of the equal legs of the triangle will be d12=1−r3 again. Similar to before, 1−r3>1−r2, so grains 1 and 2 will touch prior to the grains 1 and 3 separating. Once grains 1 and 2 touch the configuration matches the end-state configuration of the TR312 configuration. Thus the TR312 and IS31-2 configurations terminate, as shown in Fig. 4.

Stability Now consider the stability of the IS configurations. The condition δdijE≥0 is uniformly satisfied, except for the termination of the IS12-3 configuration noted above. Thus it is just needed to test whether the joint variations of δdjk and δθki are positive definite or not. For this situation one must take the second partial of the energy with respect to both of these variations, evaluated at the relative equilibrium, and test the 2 ×× 2 resulting matrix for whether it is positive definite.

$$\delta^2\mathcal{E} = \begin{bmatrix} \delta d_{jk} & \delta\theta_{ki} \end{bmatrix} \begin{bmatrix} \frac{\partial^2\mathcal{E}}{\partial d_{jk}\partial d_{jk}} & \frac{\partial^2\mathcal{E}}{\partial d_{jk}\partial\theta_{ki}} \\ \frac{\partial^2\mathcal{E}}{\partial\theta_{ki}\partial d_{jk}} & \frac{\partial^2\mathcal{E}}{\partial\theta_{jk}\partial\theta_{jk}} \end{bmatrix} \begin{bmatrix} \delta d_{jk} \\ \delta\theta_{ki} \end{bmatrix} \tag{50}$$

A matrix is positive definite by Sylvester's criterion if all of its leading principal minors are positive. A simpler, necessary condition, is that the diagonals of the matrix are all positive.

To that end, consider the term $\frac{\partial^2\mathcal{E}}{\partial\theta_{jk}\partial\theta_{jk}}$ evaluated at the equilibrium condition, which can be found to equal

$$\frac{\partial^2\mathcal{E}}{\partial\theta_{jk}\partial\theta_{jk}} = m_k m_i \left[m_k\left(m_i - 3m_j\right)d_{ki}^2 - 3I_S - 3m_i m_j(1-r_k)^2\right]\frac{\left(d_{ij}\sin\theta_{ki}\right)^2}{d_{ki}^3 I_H} \tag{51}$$

Note that the ordering of *i* and *j* does not matter, although the individual terms of the matrices may change. Thus, one can always choose to assign *i* and *j* such that mj>mi to ensure that mi−3mj<0, making the diagonal negative definite. Thus, any of the configurations will violate the necessary condition for the system to be positive definite, meaning that the Isosceles configurations are always unstable. Note that this instability mode is related to the angle θkiθki and not related to instability in the distance variation. Due to this, the IS family is always unstable even if it is formed from a Transition Fission.

Bifurcation To end, note that in Figs. 2, 4 and 5 a similar specific sequence for the evolution of all of the Isosceles configurations is shown, with them appearing as an *H*-Bifurcation with one branch continuing to ∞∞and the other terminating at a TR configuration or ending at an LO configuration. The persistence of this structure can be proven, using a similar approach as used in discussing the bifurcation of the TR132. The relation between angular momentum and distance *d* in these configurations is the simple expression H=IH(d)/d3/2. It can be shown that this function has a unique minimum positive value, and thus the isosceles configurations bifurcate into existence when the angular momentum rises above this value. Further, it can be shown that the distance at which this bifurcation occurs is always greater than the associated contact distances for this configuration. This is shown by developing a specific inequality that must be satisfied, and then checking it by computing level sets across the domain of possible radii. Doing so reveals that the bifurcation at a nonzero value of H always occurs away from any of the contact termination conditions. Thus the pattern of having one branch progress toward the TR configurations and the other branch extend to large distances can be inferred.

5.3.2 Euler Aligned Relative Equilibria

Existence The Euler Aligned relative equilibria are defined by having two grains in contact and the third at a distance along the centers of mass of the two grains in contact. Again, the grains in contact are i and j and grain k is separated. The notation for these equilibria is EA*ij-k*, where the order is important. Specifically, note that EA*ij-k* and EA*ji-k* are different, with the grain k rotated 180∘∘ relative to the other configuration. Thus, EA*ij-k* can be organized from left to right and fits with the earlier notation. There are six different configurations that can be considered, EA12-3, EA21-3, EA13-2, EA31-2, EA23-1, EA32-1. In all these definitions the angle $\delta_{\theta_{ki}}\mathcal{E} = 0.$. The two remaining conditions are then $\delta_{d_{ij}}\mathcal{E} \geq 0 \text{ and } \delta_{d_{jk}}\mathcal{E} = 0.$

For existence, solve each of these conditions for the ratio (H/IH)2 to find

$$\left(\frac{H}{I_H}\right)^2 \leq \frac{1}{m_j(1-r_k)+m_k d_{ki}}\left[\frac{m_j}{(1-r_k)^2}+\frac{m_k}{d_{ki}^2}\right] \quad (52)$$

$$\left(\frac{H}{I_H}\right)^2 = \frac{1}{m_i d_{ki}+m_j d_{jk}}\left[\frac{m_i}{d_{ki}^2}+\frac{m_j}{d_{jk}^2}\right] \quad (53)$$

Note that $d_{ki} \geq 1 + r_j \text{ and } d_{jk} = d_{ki} - (1 - r_k).$ These conditions can be combined and rewritten into a standard form

$$\frac{1}{d_{ki}^3}F(m_j/m_i, d_{jk}/d_{ki}) \leq \frac{1}{(1-r_k)^3}F(m_k/m_j, d_{ki}/(1-r_k)) \quad (54)$$

where $F(\mu, x) = (1 + \mu/x^2) / (1 + \mu x)$. Note the identity $x^3 F(\mu, x) = F(1/\mu, 1/x)$. When this inequality is violated grains *i* and *j* will separate and the configuration will cease to exist.

As a final step, define r $r = d_{ki}/(1 - r_k) > 1$, $\mu_{ij} = m_i/m_j$ and $d_{jk}/d_{ki} = 1 - 1/r$. Then the inequalities are written as

$$F(\mu_{ji}, 1 - 1/r) \leq F(\mu_{jk}, 1/r) \quad (55)$$

It can be shown (see "Appendix") that $F(\mu, r)$ is monotonically decreasing in *r* and is convex. From this it can be shown that $F(\mu, 1 - 1/r)$ is monotonically decreasing in *r* and that $F(\mu, 1/r)$ is monotonically increasing. Thus the inequality can be crossed either 0 or 1 times, and it is not needed to consider the possibility of multiple transitions in the existence of solutions. Given the well defined interval over which the parameter *r* is defined $r \in [(1 + r_j)/(1 - r_k), \infty)$, an explicit method for determining when these conditions exist can be developed.

First note that $\lim_{r \to \infty} F(\mu, 1 - 1/r) = 1$ and that $\lim_{r \to \infty} F(\mu, 1/r) \sim r^2/\mu + \ldots$. Thus the inequality is always satisfied when the distance between the grains in contact, *i* and *j*, and the separated grain *k*, is large. This holds independent of the ordering of the indices, and means that all of the EA configurations exist when the *k*th grain is sufficiently distant from the two in contact. Thus, to ascertain whether the configuration exists across all possible values of *r* it is only needed to check the condition at the minimum radius condition. Due to the topological properties of the two functions in the inequality, if the inequality is satisfied at the minimum value of *r*, then the given configuration exists across all distances in the interval. If it is violated at the minimum value of *r*, then there exists a distance at which the configuration ceases to exist.

Evaluating the inequality at the minimum distance $r = (1 + r_j)/(1 - r_k)$ yields.

$$F(\mu_{ji}, (1 - r_i)/(1 + r_j)) \leq F(\mu_{jk}, (1 - r_k)/(1 + r_j)) \quad (56)$$

If this inequality is confirmed, then the configuration EA*ij-k* exists across the whole domain and, by swapping indices *i* and *k*, that then the configuration EA*kj-i* does not exist by definition. Conversely, if the inequality for EA*ij-k* is not confirmed at the lower limit, then the configuration EA*kj-i* is trivially confirmed. The following discussions will assume that the configuration EA*ij-k* exists all the way to contact, and thus that the configuration EA*kj-i* does not and terminates at a finite distance from contact.

This means that whenever one EA configuration exists down to the Euler Resting configuration, that the alternate EA configuration does not, and terminates at a finite separation. The termination of the conjugate configuration occurs when that configuration intersects with the conditions for the orbital Euler configuration EO*ijk*, as by definition at termination $\delta_{d_{ji}}\mathcal{E} = 0$ by default and $\delta_{d_{kj}}\mathcal{E} = 0$ due to the contact constraint vanishing.

With these results in hand, the realms where the different Euler Aligned configurations exist can be evaluated. To do this, plot the level sets of the function

$$F(\mu_{jk}, (1 - r_k)/(1 + r_j)) - F(\mu_{ji}, (1 - r_i)/(1 + r_j)) = 0 \quad (57)$$

As these functions are analytical and have no singularities, there are no computational issues with evaluating these level sets. The zero line delineates where a transition in the existence of these configurations occurs. In the region where the difference is positive, the EA*ij-k* configuration exists down to contact, while in the region where the difference is negative, the EA*kj-i* configuration exists down to contact. These distinctions are important as they control which grain will separate from an Euler resting configuration when angular momentum is increased. In the following the plots of these zero lines are displayed for the different possible configurations.

EA12-3 and EA32-1 Figure 10 shows a plot of the level set of the inequality for the ordering 123, showing that there exists a region where the EA12-3 configuration exists down to the ER123 configuration, and where the EA32-1 configuration exists down to the ER123 configuration. The former exists in the region where the grains 2 and 3 are more similar sized, and the latter where the grains 1 and 2 are more similar sized. Which side of the line that configuration lies determines how the configuration will fission when it terminates. Figure 2 shows the different bifurcation pathways that occur.

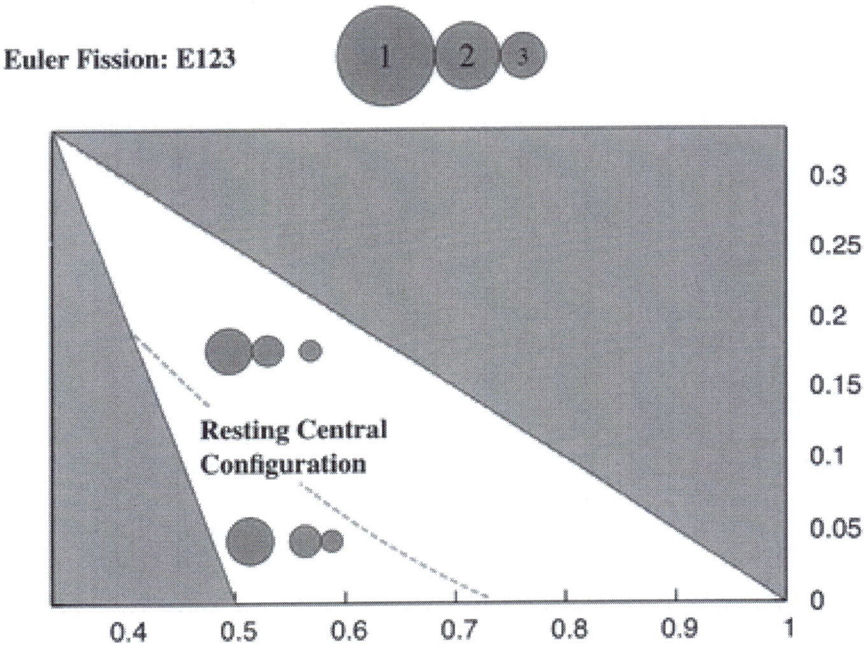

Figure 10. Fission chart for the ER123 configuration. For masses to the right of the line ER123 configuration fissions by having the smallest mass separate and transitioning into the EA12-3 configuration. For masses to the left of the line the ER123 configuration fissions by having the largest mass separate and transitioning into the EA32-1 configuration

EA13-2 and EA23-1 Figure 11 shows a plot of the level set of the inequality for the ordering 132. Here, only the EA13-2 configuration exists down to the ER132 configuration, and thus the EA23-1 configuration always terminates at a finite distance. Not shown here explicitly is that at the left border, where r1=r2r1=r2, the two conditions are equivalent due to symmetry and both EA configurations extend down to the ER132 configuration. Figure 3 shows the different bifurcation pathways that occur.

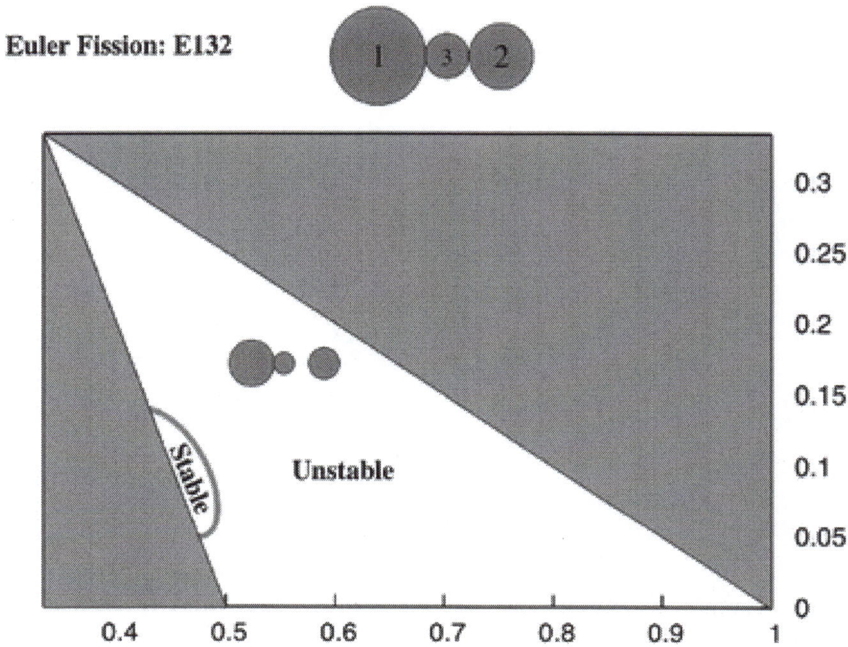

Figure 11. Fission chart for the ER132 configuration. For this configuration the intermediate mass always separates, transitioning into the EA13-2 configuration. The stability of the EA13-2 relative equilibrium at fission is shown, noting that there is a small region where the ER132 configuration fissions into a stable EA13-2 configuration

EA31-2 and EA21-3 Figure 12 shows a plot of the level set of the inequality for the ordering 312. There are two regimes again. When the grains are relatively equal in size the configuration EA21-3 continues down to ER312. Away from this geometric region, however, configuration EA31-2 continues down to ER312. Figure 4shows the different bifurcation pathways that occur.

Stability For an EA*ij-k* configuration to be stable requires δdijE>0δdijE>0 and the second variations of EE with respect to djkdjk and θkiθki be positive definite. The condition on dijdij is automatically satisfied, except at specific transition points, once it is shown that a given configuration exists. Thus only the second-order variation conditions need to be evaluated.

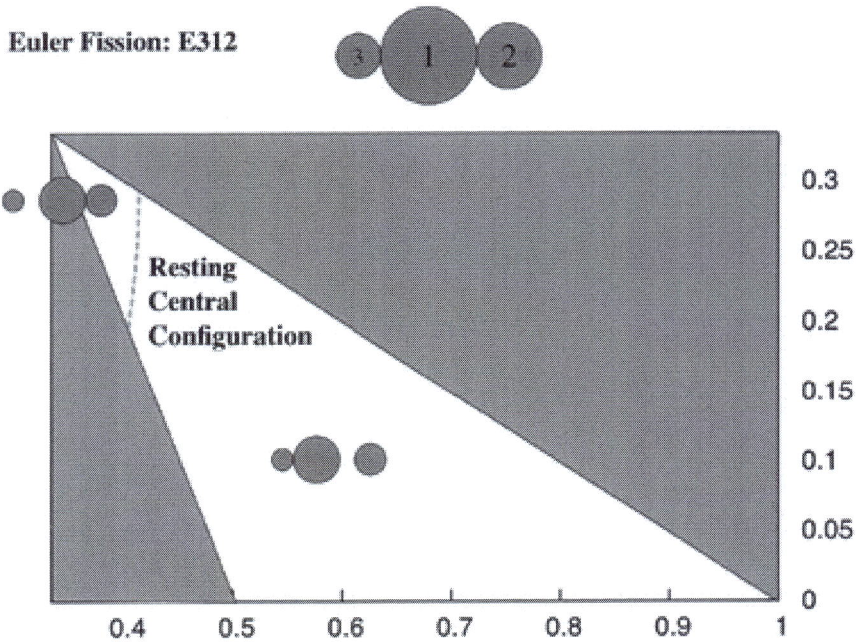

Figure 12. Fission chart for the ER312 configuration. For masses to the right of the line the ER312 configuration fissions by having the intermediate mass separate and transitioning into the EA31-2 configuration. For masses to the left of the line the ER312 configuration fissions by having the smallest mass separate and transitioning into the EA21-3 configuration

First, note that the cross variations $\delta^2_{d_{jk}\theta_{ki}}\mathcal{E}$ are identically zero, and only consider the second variations $\delta^2_{\theta_{ki}}\mathcal{E}$ and $\delta^2_{d_{jk}}\mathcal{E}$ separately. Computing the first of these and evaluating it at the nominal condition yields

$$\delta^2_{\theta_{ki}}\mathcal{E} \quad = \quad -m_k m_i \left[-\frac{H^2}{I_H^2} + \frac{1}{d_{ki}^3} \right] (1 - r_k) \left(d_{ki} - (1 - r_k) \right) (\delta\theta_{ki})^2 \tag{58}$$

Make the substitution H2/ $H^2/I_H^2 = \frac{1}{d_{ki}^3} F(m_j/m_i, 1 - (1 - r_k)/d_{ki})$.
Then, stability in this variation can be established by showing that

$$F(m_j/m_i, 1-(1-r_k)/d_{ki}) \;>\; 1 \tag{59}$$

However, the function $F(\mu, 1-1/r)$ was shown to be monotonically decreasing in r with the limiting value of 1 as r becomes arbitrarily large. Thus this is always satisfied and the EA configurations are always stable to variations in the angle θ_{ki}.

All that is left is to consider when $\delta^2_{d_{jk}}\mathcal{E} > 0.$ First, re-express Eq. (87) as

$$\delta_{d_{jk}}\mathcal{E} \;=\; \frac{m_k}{I_H^2}\left\{-H^2\left(m_j+m_i\right)+I_H^2\left[\frac{m_j}{d_{jk}^2}+m_i\frac{d_{jk}-d_{ij}\cos\theta_{ki}}{d_{ki}^3}\right]\right\}\delta d_{jk}$$

(60)

The term inside the brackets is identically zero at equilibrium, thus one does not need to take the variation of terms outside of the brackets. Taking the variation inside the brackets and simplifying yields

$$\delta^2_{d_{jk}}\mathcal{E} \;=\; 2m_k\left\{\frac{2m_k}{I_H}\left(m_jd_{jk}+m_id_{ki}\right)\left[\frac{m_j}{d_{jk}^2}+\frac{m_i}{d_{ki}^2}\right]-\left[\frac{m_j}{d_{jk}^3}+\frac{m_i}{d_{ki}^3}\right]\right\}\delta d^2_{jk}$$

(61)

First consider the case when $d \sim d_{jk} \sim d_{ki} \gg 1.$ The second variation then reduces to

$$\delta^2_{d_{jk}}\mathcal{E} \;=\; 2m_k\frac{m_j+m_i}{d^3}\delta d^2_{jk} \tag{62}$$

which is always positive. Thus, all EA*ij*-*k* configurations with large enough distances are stable.

The stability of the ER *ijk* configurations at fission can be checked using Eq. (61) by evaluating it at the contact conditions and plotting level sets of its zero value. Doing so shows that configurations ER123 and ER312 are always unstable when they fission. The configuration ER132 is found to have a small region where it is stable when it fissions, in the vicinity of the region which had stable TR132 configurations, and shown in Fig. 11. Over the remaining domain the ER132 configuration is unstable when it fissions. From Corollary 1 we conclude that the region where the ER132 is stable at fission corresponds to a Transition Fission (see Fig. 3, bottom). From the uniqueness of the Euler Orbital configurations these EA13-2 configurations should remain stable for all larger distances. Conversely, for the unstable fission cases, Corollary 1 implies that these are Termination Fissions. For these cases, consider the stability of their EA configurations as a function of separation. At their minimum separation, when terminating the ER configurations, they begin as unstable. As the distance is increased they eventually become stable, indicating that EdjkdjkEdjkdjk evaluated at the EA configuration must cross through zero at some specific equilibrium configuration, indicating the point where the *H*-Bifurcation occurs.

Bifurcation Studying this aspect of the situation can provide qualitative insight into how the EA configurations bifurcate into existence and evolve as *H* is increased. Consider Eq. (53), which must be satisfied for an EA relative equilibrium configuration. As all of the terms on the right hand side are positive and bounded from below, there is an absolute minimum value such that if *H* is below this value the equality cannot be satisfied and the EA relative configuration does not exist. At this value of angular momentum there will be a bifurcation from no relative equilibria to two relative equilibria, corresponding to the point identified above where $\mathcal{E}_{d_{jk}d_{jk}} = 0.$ The exception are the stable ER132 fissions, where the bifurcation point occurs at a closer distance not allowed. For the other cases, as *H* increases further one branch of the EA relative equilibria must migrate toward the ER configuration and the other to larger distances, due to the uniqueness of this family. The branch that migrates to the larger separation will be stable while the branch that migrates to the contact configuration must be unstable, from Lemma 5. A similar bifurcation will occur for the configuration that intersects with the Euler Orbital family.

5.4 No Active Constraint

Finally consider the case where none of the constraints are active. Then the three conditions must all be identically zero. Due to the structure of the problem, it is well known that there are only five relative equilibria to this problem (Wintner 1941). These are divided into the Lagrange solutions, which lie at the vertices of an equilateral triangle (Lagrange 1772), and the Euler solutions, which lie in a single line and are appropriately spaced (Euler 1767).

5.4.1 Lagrange Solutions

For the Lagrange solutions, set $d = d_{12} = d_{23} = d_{31}$. From Eq. (3) note that this requires cosθ31=1/2cos⁡θ31=1/2, meaning that $\theta_{31} = \pm 60^\circ$. Then the condition can be uniformly satisfied by choosing the distance *d* such that $H^2 = \frac{I_H(d)^2}{d^3}$. Note that for all $d > \max(r_i + r_j) = r_1 + r_2$ (given the assumed ordering) such a solution will always exist. However, for a given H2 a solution to the non-contact case may not always exist. Indeed, since for this case $I_H = (m_1 m_2 + m_2 m_3 + m_3 m_1)d^2 + I_S$, H has a minimum value that is greater than zero, and thus will not exist for all values of angular momentum.

The point where the Lagrange orbital configuration comes into existence can be explicitly probed. In general there are two possibilities. One is that it appears as a two branch family once the angular momentum goes above its minimum value. Then one branch will migrate inwards with increasing angular momentum and terminate by intersection with the IS12-3 family. Otherwise, if the minimum angular momentum point arises at a mutual distance less than r1+r2r1+r2, then the inner IS12-3 family will transition directly into the LO family, as it is known that the equal mass case has this sort of a bifurcation (Scheeres 2012), it is relevant to test for when this will occur. To do this just compute ∂H/∂d and solve for the zero to find

$$d^2 = \frac{3 I_S}{m_1 m_2 + m_2 m_3 + m_3 m_1} \tag{63}$$

Which bifurcation structure ensues can be found by finding where this solution is greater or less than (1−r3). Figure 13 plots the region where the double branch occurs and where the single branch occurs. Figure 5 shows the two different bifurcation pathways.

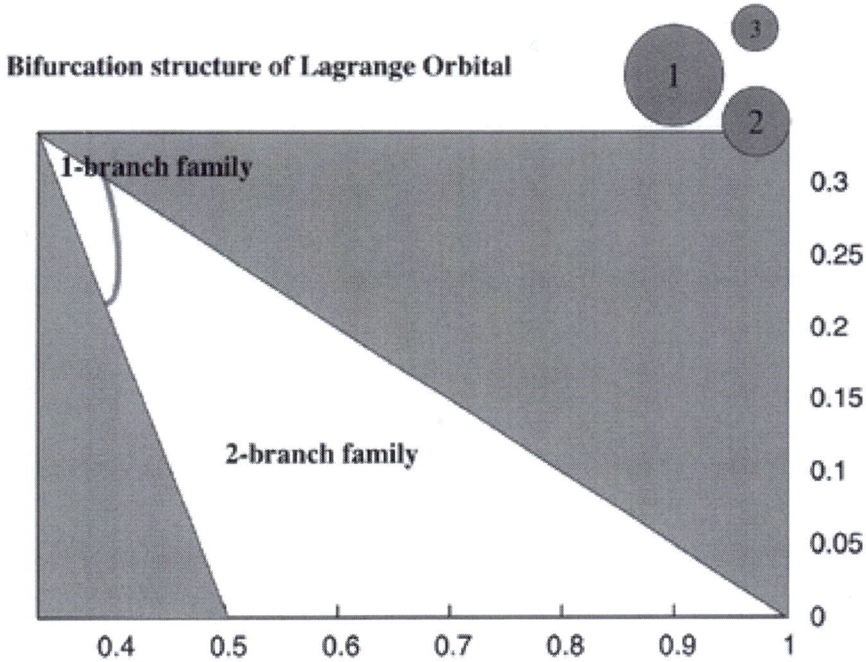

Figure 13. Transition line between the two different bifurcation modes of the Lagrange orbital configuration. The two different patterns of bifurcation are shown in Fig. 5

While it is well known that the classical three-body problem is spectrally stable when the Routh Criterion is satisfied, it should be noted that the stability considered in this paper, energetic stability, is a stronger type of stability. An observation by Moeckel (1990) shows that central configurations in the point-mass *N*-body problem never have a positive definite second variation of their energy, and thus it can be suspected that the same holds true for the Lagrange orbital configuration in the full body. To test this, take the second-order variation of E, evaluated at the equilibrium, and determine if the resulting matrix of values is positive definite. Here it is simpler to take the 3 distances d12, d23 and d31 as the degrees of freedom, with the general form, starting from Eq. (89), substituting the equilibrium condition and simplifying

$$\frac{\partial^2 \mathcal{E}}{\partial d_{ij}^2} = \frac{m_i m_j}{d I_H}\left[\left(m_i m_j - 3m_k(m_i + m_j)\right)d^2 - 3I_S\right] \quad (64)$$

$$\frac{\partial^2 \mathcal{E}}{\partial d_{ij}\partial d_{jk}} = \frac{4m_i m_j^2 m_k}{d I_H} \quad (65)$$

where d≥1−r3 for the specific case of interest.

For the full Hessian of E, [∂2E/∂dijdjk], to be positive definite utilize Sylvester's Theorem again, which states that a necessary and sufficient condition is that all of the principal minors of the Hessian matrix be positive. Thus, a necessary condition for being positive definite is that the diagonals all be positive. Should any of these be negative, then the matrix is not positive definite and hence the relative equilibrium configuration is not energetically stable. Consider the entry for $\frac{\partial^2 \mathcal{E}}{\partial d_{23}^2}$. The controlling condition for stability is then that $\left[m_2 m_3 - 3m_1(m_2 + m_3)\right]d^2 - 3I_S$ be positive. However, it is easy to show that the term $\left[m_2 m_3 - 3m_1(m_2 + m_3)\right] < 0,$ showing that the Lagrange orbital configurations are always energetically unstable, consistent with Moeckel's result. First, restate the negative condition as $3m_1(m_2 + m_3) > m_2 m_3,$ then note that $m_2 + m_3 > m_3 \text{ and } m_1 > m_2,$ establishing the inequality unequivocally.

5.4.2 Euler Solutions

For the Euler conditions consider $d_{ij} \geq r_i + r_j,\ d_{jk} \geq r_j + r_k, \text{ and } \theta_{ki} = \pi.$ This case also has $d_{ki} = d_{ij} + d_{jk}.$. Both Eqs. (86) and (87) must equal zero in this case, yielding the two conditions.

$$0 = m_i m_j \left[-\frac{H^2}{I_H^2} + \frac{1}{d_{ij}^3} \right] d_{ij} + m_i m_k \left[-\frac{H^2}{I_H^2} + \frac{1}{d_{ki}^3} \right] d_{ki} \quad (66)$$

$$0 = m_k m_j \left[-\frac{H^2}{I_H^2} + \frac{1}{d_{jk}^3} \right] d_{jk} + m_k m_i \left[-\frac{H^2}{I_H^2} + \frac{1}{d_{ki}^3} \right] d_{ki} \quad (67)$$

First, there is a more fundamental equality within these results

$$m_i m_j \left[-\frac{H^2}{I_H^2} + \frac{1}{d_{ij}^3} \right] d_{ij} = m_k m_i \left[\frac{H^2}{I_H^2} - \frac{1}{d_{ki}^3} \right] d_{ki} = m_k m_j \left[-\frac{H^2}{I_H^2} + \frac{1}{d_{jk}^3} \right] d_{jk} \quad (68)$$

By inspection, with the knowledge that dki≥djk,dijdki≥djk,dij, note that

$$\frac{1}{d_{ki}^3} \leq \frac{H^2}{I_H^2} \leq \frac{1}{\max\left(d_{jk}, d_{ij}\right)^3} \quad (69)$$

Alternately, this ratio can also be solved for the quantity (H/IH)2(H/IH)2 to find

$$\frac{H^2}{I_H^2} = \frac{1}{m_j d_{ij} + m_k d_{ki}} \left[\frac{m_j}{d_{ij}^2} + \frac{m_k}{d_{ki}^2} \right] \quad (70)$$

$$\frac{H^2}{I_H^2} = \frac{1}{m_j d_{jk} + m_i d_{ki}} \left[\frac{m_j}{d_{jk}^2} + \frac{m_i}{d_{ki}^2} \right] \quad (71)$$

which is the condition used to analyze how the ER configurations fissioned. Indeed, at the transition lines on Figs. 10 and 12 the resting configuration is in fact a central configuration, meaning that the relative attractions between the bodies will be balanced so long as their relative distances are preserved.

This can be generalized to identify the possible bifurcation pattern in the EO configurations. Assume, say, that bodies *i* and *j* are in contact and that as body *k* is moved to a larger distance it reaches the point where the equality between the above conditions occurs, meaning that bodies *i* and *j* are now in a relative equilibrium condition and the entire system satisfies a central configuration conditions (Wintner 1941). At this point the relative distances between these bodies can be uniformly scaled with the ratio **H/IH** following along. The change in angular momentum with this scaling is not uniform, however, due to the 3IS3IS term in IH. Of specific interest regarding the pattern of bifurcation is whether the angular momentum decreases or increases with this change in relative distance. In the following it can be shown that both conditions can occur in general.

Define the distance between bodies *k* and *i* where the EO conditions are satisfied
$(\text{assuming } d_{ij} = 1 - r_k)$ as d^*_{ki}, and thus $d^*_{jk} = d^*_{ki} - (1 - r_k)$, define the ratio $H/I_H = \Omega^*$ at this point. Then, for increasing the distance the relative equilibria will all scale uniformly, meaning that

$$\frac{H^2}{I_H^2} = \frac{\Omega^{*2}}{d^3} \tag{72}$$

where $d \geq 1$ and $d_{ij} = d(1 - r_k)$, $d_{jk} = dd^*_{jk}$ and $d_{ki} = dd^*_{ki}$. With this structure, it is possible to compute the gradient

$$\frac{\partial H}{\partial d} = \frac{\Omega^{*2}}{2d^{5/2}} \left[\left(m_i m_j (1 - r_k)^2 + m_j m_k (d^*_{ki} - (1 - r_k))^2 + m_k m_i d^{*2}_{ki} \right) d^2 - 3 \right] \tag{73}$$

How the bifurcation occurs can be tested by plotting the level sets from $d^*_{ki} = (1 + r_j)$ to large values. For the 1, 2, 3 and 2, 1, 3 orderings the gradient is positive toward the apex of the triangular region and can take on negative values near the base. Thus, the appearance of the EO orbits occur as a transition closer to the equal mass condition and as a bifurcation followed by a termination away from there. Precise limits could be computed, but would require root solving algorithms.

Finally, consider the stability of the Euler Orbital solutions. These are again suspected to be energetically unstable due to the instability of the point-mass cases; however, this should be checked given the changes in the current approach. First, note that the second-order variation in θkiθki is uncoupled from the variations in distance, and evaluated at the equilibrium yields

$$\delta^2_{\theta_{ki}}\mathcal{E} = -m_k m_i \left[-\frac{H^2}{I_H^2} + \frac{1}{d_{ki}^3}\right] d_{ij} d_{jk} (\delta\theta_{ki})^2 \quad (74)$$

The quantity in the brackets is negative, as established above, and hence the angle variation is stable. For the distance variations the full 2×22×2 Hessian matrix must be evaluated, however one can again just check the necessary conditions that the diagonals must all be positive. Taking the second-order variations of both conditions from the diagonal of the Hessian matrix shows that both of the following conditions must be positive for stability

$$m_i m_j - 3(m_i + m_j)m_k > 0 \quad (75)$$

$$m_j m_k - 3(m_j + m_k)m_i > 0 \quad (76)$$

For the ordering $m_3 \leq m_2 \leq m_1$ it can be shown that for all combinations of *i*, *j* and *k* that at least one of these conditions will be violated, and hence the EO configurations are always unstable.

6 SUMMARY

To finish, the results are presented in light of the main theorem. First, consider the total number of relative equilibria found. For the no contact case recall the classical result of five distinct relative equilibria. When one contact is active the EA and IS relative equilibria were identified, which have 6 unique components each, raising the count to 17. For the two contact cases there are three ER and six TR configurations, resulting in a total of 26. Finally, for three contact cases there are the two LR configurations, leading to the total of 28.

Now consider the bifurcation patterns, which are focused on the transitions between the different contact cases, and the identification of when the *H*-Bifurcations can occur. The details of the transitions will be outlined, although a few observations can be given first. With regard to stability the system starts with only two stable LR configurations at low values of *H* and eventually has six stable EA configurations for arbitrarily large values of *H*. Between these limits the number of stable configurations can vary, and to establish the precise sequence and number would require a more detailed investigation for a specific set of sizes. It is noted, however, that there always exist at least one stable configuration.

The bifurcation pattern seen in Fig. 3 is described first. The LR configurations all end at a Termination Fission condition with the TR132 configurations. These configurations either arise from an *H*-Bifurcation (in a limited region of the parameter space) or more commonly emerge as a symmetric bifurcation as the unstable ER132 configuration stabilizes. Under increasing *H* the ER132 configuration either ends with a Termination Fission with the unstable component of the EA13-2 configuration, or for a limited range of parameters ends with a Transition Fission into the EA13-2 configuration. This second occurrence is of great interest as it is the only occasion in which the spherical full three-body problem will fission into a stable configuration. The EA13-2 configuration itself usually arises as an *H*-Bifurcation, with its unstable branch terminating as mentioned above and its stable branch existing for all *H* with an increasing distance proportional to H2H2. The only exception is when it arises as a Transition Fission, as described above. The EA23-1 configuration has an evolution that is completely isolated from the rest of this chart. It arises through an *H*-Bifurcation, with its unstable component either having a Termination Fission with the EO132 configuration or a Transition Fission into an EO132 configuration. Its stable component continues for all larger values of *H* with a similar asymptotic form as the EA13-2 configuration. The EO132 configuration can either arise as an *H*-Bifurcation or through a Transition Fission, however the EO components are always unstable. It is important to note that at each stage of the system evolution with *H*, that there is at least one stable configuration, providing proof of that aspect of the theorem.

Now consider the bifurcation patterns for the ER123 and ER213 pathways, shown in Figs. 2 and 4. These are similar, and distinct from the ER132 pathway. Each of these start out in unstable ER configurations and both stabilize by a symmetric bifurcation with the TR123 and TR213 configurations, respectively. The TR123 and TR213 configurations end with a Termination Fission with an IS23-1 and IS13-2 configuration, respectively. The IS configurations arise through an *H*-Bifurcation with the inner configuration ending with the Termination Fission mentioned above and the outer configuration extending for all *H*, ultimately with their size on the order of H2, although the IS configurations are always unstable. The stable ER123 and ER213 configurations end with a Termination Fission into an EA12-3 or EA32-1 configuration for the ER123 case or an EA21-3 or EA31-2 configuration for the ER213 case. Limits where these transitions occur have been delineated in Figs. 10 and 12. Note that at the transition between these fission pattern the ER123 and ER213 configurations are in a central configuration, a situation that does not happen for the ER132 configuration. When the EA inner configurations do not end with a Termination Fission with an ER configuration, they end with a Termination Fission or Transition Fission with the corresponding EO configuration. The outer EA configurations are all stable and have the same asymptotic structure for large *H*. The EO123 and EO213 configurations either arise as an *H*-Bifurcation or as a Transition Fission, with the details of these boundaries left for future investigation.

Finally consider the sequence involving the LO configuration, represented in Fig. 5. This sequence is the least complex, with the IS12-3 configurations arising from an *H*-Bifurcation. The inner component ends with a Termination Fission or Transition Fission with the LO configuration. The LO configuration, in turn, either arises as an *H*-Bifurcation, with the inner component ending with a Termination Fission, or as a Transition Fission. All configurations in these sequences are unstable.

ACKNOWLEDGMENTS

The author acknowledges support from NASA grant NNX14AL16G from the Near Earth Objects Observation programs.

APPENDIX

Partial Derivatives

It is useful to state the relevant partial derivatives of the amended potential and its constituent terms, as a function of the distances and angles. In the following use the convention that the distances are denoted with indices *ij* and *jk* and the angle with indices *ki*.

If the third degree of freedom is the angle θkiθki then

$$\frac{\partial \mathcal{E}}{\partial d_{ij}} = -\frac{H^2}{2I_H^2}\frac{\partial I_H}{\partial d_{ij}} + \frac{\partial \mathcal{U}}{\partial d_{ij}} \tag{77}$$

$$\frac{\partial \mathcal{E}}{\partial \theta_{ki}} = -\frac{H^2}{2I_H^2}\frac{\partial I_H}{\partial \theta_{ki}} + \frac{\partial \mathcal{U}}{\partial \theta_{ki}} \tag{78}$$

where

$$\frac{\partial I_H}{\partial d_{ij}} = 2m_i m_j d_{ij} + 2m_i m_k \left(d_{ij} - d_{jk}\cos\theta_{ki}\right) \tag{79}$$

$$\frac{\partial I_H}{\partial \theta_{ki}} = 2m_i m_k d_{ij} d_{jk}\sin\theta_{ki} \tag{80}$$

Similarly

$$\frac{\partial \mathcal{U}}{\partial d_{ij}} = m_i m_j \frac{1}{d_{ij}^2} + m_i m_k \frac{\left(d_{ij} - d_{jk}\cos\theta_{ki}\right)}{d_{ki}^3} \tag{81}$$

$$\frac{\partial \mathcal{U}}{\partial \theta_{ki}} = m_i m_k \frac{d_{ij} d_{jk}\sin\theta_{ki}}{d_{ij}^3} \tag{82}$$

If the third degree of freedom is the distance dkidki, and if not at a limiting constraint, then

$$\frac{\partial \mathcal{E}}{\partial d_{ij}} = -\frac{H^2}{2I_H^2}\frac{\partial I_H}{\partial d_{ij}} + \frac{\partial \mathcal{U}}{\partial d_{ij}} \tag{83}$$

where

$$\frac{\partial I_H}{\partial d_{ij}} = 2m_i m_j d_{ij} \tag{84}$$

and

$$\frac{\partial \mathcal{U}}{\partial d_{ij}} = m_i m_j \frac{1}{d_{ij}^2} \tag{85}$$

Equilibrium Conditions

For a relative equilibrium there are two different possibilities. Either δE=0 or δE>0. For either, the relevant statement of the variations is given in the following for the two different formulations of the third degree of freedom.

If the third degree of freedom is the angle θkiθki, then the full set of variations are

$$\begin{aligned}\delta_{d_{ij}}\mathcal{E} &= m_i\left\{m_j\left[-\frac{H^2}{I_H^2}+\frac{1}{d_{ij}^3}\right]d_{ij}\right.\\ &\left.+m_k\left[-\frac{H^2}{I_H^2}+\frac{1}{d_{ki}^3}\right][d_{ij}-d_{jk}\cos\theta_{ki}]\right\}\delta d_{ij}\end{aligned} \tag{86}$$

$$\begin{aligned}\delta_{d_{jk}}\mathcal{E} &= m_k\left\{m_j\left[-\frac{H^2}{I_H^2}+\frac{1}{d_{jk}^3}\right]d_{jk}\right.\\ &\left.+\,m_i\left[-\frac{H^2}{I_H^2}+\frac{1}{d_{ki}^3}\right][d_{jk}-d_{ij}\cos\theta_{ki}]\right\}\delta d_{jk}\end{aligned} \tag{87}$$

$$\delta_{\theta_{ki}}\mathcal{E} = m_k m_i\left[-\frac{H^2}{I_H^2}+\frac{1}{d_{ki}^3}\right]d_{ij}d_{jk}\sin\theta_{ki}\delta\theta_{ki} \tag{88}$$

If the third degree of freedom is the distance d_{ki}, and the system is not at a constraint limit then the full set of variations are

$$\delta_{d_{ij}}\mathcal{E} = m_i m_j\left[-\frac{H^2}{I_H^2}+\frac{1}{d_{ij}^3}\right]d_{ij}\delta d_{ij} \tag{89}$$

$$\delta_{d_{jk}}\mathcal{E} = m_j m_k\left[-\frac{H^2}{I_H^2}+\frac{1}{d_{jk}^3}\right]d_{jk}\delta d_{jk} \tag{90}$$

$$\delta_{d_{ki}}\mathcal{E} = m_k m_i \left[-\frac{H^2}{I_H^2} + \frac{1}{d_{ki}^3} \right] d_{ki}\delta d_{ki} \tag{91}$$

These expressions are used to develop the necessary and sufficient conditions for a configuration to be a relative equilibrium. If any two components are in contact, then the condition for the degree of freedom that is blocked should be δE>0 for all allowed variations, otherwise the condition should be δE=0δ

Properties of the Function F(μ,x)

Lemma 7

The function

$$F(\mu, x) = \frac{1 + \frac{\mu}{x^2}}{1 + \mu x} \tag{92}$$

is monotonically decreasing in *x* and is convex in *x* over the interval x∈(0,∞)x∈(0,∞).

Proof

Consider the first derivative of the function with respect to *x*:

$$F'(\mu, x) = \frac{-2\mu}{x^3}\frac{1}{1+\mu x} - \mu\frac{1 + \frac{\mu}{x^2}}{(1+\mu x)^2} \tag{93}$$

By inspection it can be seen that all terms are negative and nonzero, and thus the function is monotonically decreasing in *x*.

Taking the second derivative with respect to *x*:

$$F''(\mu, x) = \frac{6\mu}{x^4}\frac{1}{1+\mu x} + \frac{4\mu^2}{x^3}\frac{1}{(1+\mu x)^2} + 2\mu^2\frac{1 + \frac{\mu}{x^2}}{(1+\mu x)^3} \tag{94}$$

By inspection again see that all terms are positive and nonzero and is thus convex.

REFERENCES

1. Arnold, V.I., Kozlov, V.V., Neishtadt, A.I.: Encyclopaedia of Mathematical Sciences: Dynamical Systems III. Springer, Berlin (1988)

2. Euler, L.: De motu rectilineo trium corporum se mutuo attrahentium. Novi Comm. Acad. Sci. Imp. Petrop. **11**, 144–151 (1767)

3. Fujiwara, A., Kawaguchi, J., Yeomans, D.K., Abe, M., Mukai, T., Okada, T., Saito, J., Yano, H., Yoshikawa, M., Scheeres, D.J., et al.: The rubble-pile asteroid Itokawa as observed by Hayabusa. Science **312**(5778), 1330–1334 (2006)

4. Greenwood, D.T.: Principles of Dynamics, 2nd edn. Prentice-Hall, Englewood Cliffs, NJ (1988)

5. Lagrange, J.-L.: Essai sur le probleme des trois corps. Prix de lAcadémie royale des Sciences de Paris **9**, 229–324 (1772)

6. Massironi, M., Simioni, E., Marzari, F., Cremonese, G., Giacomini, L., Pajola, M., Jorda, L., Naletto, G., Lowry, S., El-Maarry, M.R., et al.: Two independent and primitive envelopes of the bilobate nucleus of comet 67P. Nature **526**(7573), 402–405 (2015)

7. Moeckel, R.: On central configurations. Mathematische Zeitschrift **205**(1), 499–517 (1990)

8. Richardson, D.C., Elankumaran, P., Sanderson, R.E.: Numerical experiments with rubble piles: equilibrium shapes and spins. Icarus **173**(2), 349–361 (2005)

9. nchez, P., Scheeres, D.J.: DEM simulation of rotation-induced reshaping and disruption of rubble-pile asteroids. Icarus **218**, 876–894 (2012)

10. Scheeres, D.J.: Stability in the full two-body problem. Celest. Mech. Dyn. Astron. **83**(1), 155–169 (2002)

11. Scheeres, D.J.: Minimum energy configurations in the n-body problem and the celestial mechanics of granular systems. Celest. Mech. Dyn. Astron. **113**(3), 291–320 (2012)

12. Scheeres, D.J.: Hill stability in the full 3-body problem. Proc. Int. Astron. Union **9**(S310), 134–137 (2014)

13. Scheeres, D.J.: Hill Stability of Configurations in the Full NN-Body Problem. In: Proceedings of the International Astronomical Union (S318) (2015)

14. Scheeres, D.J.: Relative equilibria in the full n-body problem with applications to the equal mass problem. In: Bonnard, B., Chyba, M. (eds.) Recent Advances in Celestial and Space Mechanics, Mathematics for Industry, vol. 23, pp. 31–81 (2016)

15. Smale, S.: Topology and mechanics. I. Inventiones Mathematicae **10**(4), 305–331 (1970)

16. Smale, S.: Topology and mechanics. II. Inventiones Mathematicae **11**(1), 45–64 (1970)

17. Wintner, A.: The Analytical Foundations of Celestial Mechanics. Princeton University Press, Princeton (1941)

4

CHAPTER

Computing (Un)Stable Manifolds With Validated Error Bounds: Non-Resonant and Resonant Spectra

Jan Bouwe van den Berg

Department of Mathematics, VU University Amsterdam , Jason D. Mireles James , Christian Reinhardt

ABSTRACT

We develop techniques for computing the (un)stable manifold at a hyperbolic equilibrium of an analytic vector field. Our approach is based on the so-called parametrization method for invariant manifolds. A feature of this approach is that it leads to a posteriori analysis of truncation errors which, when combined with careful management of round off errors, yields a mathematically rigorous enclosure of the manifold. The main novelty of the present work is that, by conjugating the dynamics on the manifold to a polynomial rather than a linear vector field, the computer-assisted analysis is successful even in the case when the eigenvalues fail to satisfy non-resonance conditions. This generically occurs in parametrized families of vector fields. As an example, we use the method as a crucial ingredient in a computational existence proof of a connecting orbit in an amplitude equation related to a pattern formation model that features eigenvalue resonances.

Keywords: Computer-assisted proof Invariant manifolds Parametrization method Resonances Contraction mapping

1 INTRODUCTION

Stable and unstable manifolds are fundamental building blocks for understanding the global dynamics of nonlinear differential equations. Since closed-form analytic expressions for stable/unstable manifolds are rarely available, considerable effort goes into developing numerical techniques for their approximation, see e.g., Krauskopf et al. (2005), Haro and da la Llave (2006), Beyn and Kless (1998) and Castelli et al. (2015) and the references therein. One powerful tool for studying invariant manifolds (stable, unstable, strongly (un)stable and even more general ones) is the parameterization method of Cabré et al. (2003a, b, 2005). The parameterization method is based on formulating certain operator equations (invariance equations) which simultaneously describe both the dynamics on the manifold and its embedding. The method has been implemented numerically to study a variety of problems involving stable and unstable manifolds of equilibria and fixed points (Mireles-James 2013; Mireles-James and Mischaikow 2013; Mireles James and Lomelí2010; Haro 2011; van den Berg et al. 2011), stable/unstable manifolds of periodic orbits for differential equations (Guillamon and Huguet 2009; Huguet and de la Llave 2013; Castelli et al. 2015), quasi-periodic invariant sets in dynamical systems (Haro and da la Llave 2006) and more recently in order to simultaneously compute invariant manifolds with their unknown dynamics (Canadell and Haro 2014; Haro et al. 2014), to mention just a few examples.

The last reference is a book which provides many other examples and much fuller discussion of the literature.

In addition to facilitating efficient numerical computations, the functional analytic framework of the parameterization method also provides a natural setting for a posteriori analysis of errors. The works of van den Berg et al. (2011), Mireles-James and Mischaikow (2013) and Mireles-James (2015) exploit this a posteriori analysis and implement mathematically rigorous numerical validation methods for the stable and unstable manifolds. The term "validation" here expresses the fact that the computations provide explicit bounds on all approximation errors involved.

Approximate parametrizations are often computed by substituting a power series ansatz into the invariance equation and deriving a sequence of *homological equations* for the power series coefficients.

These homological equations are then solved *recursively* to any desired order. In this paper we employ an alternative methodology for solving the invariance equation. We recast the infinite system of homological equations as a nonlinear zero finding problem on a Banach space of geometrically decaying sequences, and we implement a parametrized Newton–Kantorovich argument in the style of Yamamoto (1998).

There are three advantages to using a Newton method. First, we note the works of van den Berg et al. (2011), Mireles-James and Mischaikow (2013) and Mireles-James (2015) assume that certain non-resonance conditions between the eigenvalues are satisfied. The main goal of the present work is to weaken this assumption. We build on the theory of Cabré et al. (2003a, b, 2005) and develop computer-assisted methods for rigorous error bounding even in the face of a resonance. More precisely we develop validation schemes which apply at any co-dimension one resonance between the eigenvalues. The formulation of the resonant as well as non-resonant zero finding problems on an infinite sequence space unifies the presentation and implementation as well as the necessary a posteriori analysis.

Second, the zero finding methodology based on a numerical Newton method for the truncated problem leads to improved numerical performance even in the case where validated numerics are not desired. The Newton iteration can always be started from the linear approximation of the manifold by its eigenvectors; however, one has the option of improving the convergence by starting the iteration from a polynomial obtained by solving a few of the lower-order homological equations recursively. Once the iteration begins the order of the polynomial approximation is roughly doubled at each step. The freedom one has in choosing the lengths of the eigenvectors is exploited in order to guarantee that the Taylor coefficients decay at the desired rate. A similar numerical Newton scheme for computing invariant manifolds in discrete time dynamical systems (without rigorous validation) was used in Mireles James and Lomelí (2010). Indeed this kind of Newton scheme is always needed when the parameterization method is applied in Fourier space, see again Haro et al. (2014) and the references therein.

Third, existing rigorous continuation methods (van den Berg et al. 2010; Breden et al. 2013) are directly applicable to our novel approach, thus putting the rigorous computation of branches of connecting orbits within direct reach. Of course, from the viewpoint of continuation, all three advantages are crucial. Indeed, while continuing along one-parameter families of (un)stable manifolds resonances are encountered generically and are thus unavoidable.

Our use of parameterized Newton–Kantorovich arguments is guided by the work of a number of authors on the so-called method of radii polynomials. This method provides a tool kit for solution and continuation of zero finding problems in infinite dimensions. In this method one derives certain polynomials whose coefficients encode information about the approximate solution, the choice of approximate inverse of the derivative, the local regularity properties of the problem and the choice of Banach space on which to work. Once these givens are fixed the roots of the radii polynomials yield not only existence and uniqueness results, but also tight error estimates and isolation bounds for the problem. Radii polynomial methods have been applied successfully to a number of problems in dynamical systems, partial differential equations and delay equations, and we refer the interested reader to Day et al. (2007), Breden et al. (2013), van den Berg et al. (2010), van den Berg et al. (2011), Hungria et al. (2016) and van den Berg et al. (2015) for more discussion and references.

Let us now be more explicit about the above-mentioned resonance conditions (see Sects. 2.2, 2.3 and 2.4and Cabré et al. 2003a, b, 2005 for full details). We consider a stationary point of an *analytic* vector field u′=g(u), with u∈Rn. Focusing on the stable manifold, we denote the eigenvalues with negative real parts by λd, where *d* is the dimension of the manifold. A *resonance* is a non-trivial relation

$$\tilde{k} \cdot \lambda \stackrel{\text{def}}{=} \tilde{k}_1\lambda_1 + \tilde{k}_2\lambda_2 + \cdots + \tilde{k}_d\lambda_d = \lambda_{\tilde{\imath}},$$

for some $\tilde{\imath} \in \{1, \ldots, d\}$ and $\tilde{k}_i \in \mathbb{N}$ for $i = 1, \ldots, d$. The trivial, excluded case is $\tilde{k} = e_{\tilde{\imath}}$, where we use the notation eiei for the *i*-th unit vector. If there are no resonances between the stable eigenvalues (the “non-resonant” case), then there is an analytic conjugacy between the flow of u′=g(u)u′=g(u) on the local stable manifold and the linear flow $\theta_i' = \lambda_i\theta_i,\ i = 1, \ldots, d.$. Since the conjugacy map $u = P(\theta)$ is analytic, *P*can be written as a convergent power series in θ.

When there is a resonance between the stable eigenvalues, the conjugacy map *P* as constructed above is no longer analytic and cannot be expressed as a convergent power series. One way to resolve this obstacle is to change the flow θ′=h(θ) in parameter space to a nonlinear "normal form," in such a way that the conjugacy is again analytic. In this paper we consider two types of resonances in particular, namely the co-dimension *one* resonances. The first type of resonance is a single "regular" resonance $\tilde{k} \cdot \lambda = \lambda_{\tilde{i}} \in \mathbb{R}$ for some $1 \leq \tilde{i} \leq d$ and $\tilde{k} \in \mathbb{N}^d$ with $\sum_{i=1}^{a} k_i \geq 2$, and no other resonances. The second type is a double real eigenvalue (k~=eik~=ei for some i≠ı~i≠ı~) with geometric multiplicity one, and no other resonances. We note that a double eigenvalue is not a resonance in the strict sense, but we nevertheless use this "uniform" terminology in the current paper. These are the only co-dimension one resonances; hence, they are the types that are encountered generically in one-parameter continuation. For this reason we restrict our attention to these resonance types as our examples. We note that a completely analogous approach works for resonances of higher co-dimension, but here omit the details.

In order to illustrate the application of our methods we discuss three example problems in detail. First we analyze the stable manifold of the origin in the well-known Lorenz equations. We use this model system to scrutinize our method in the non-resonant case and show how the structure of the vector field is directly reflected in the bounds used for validation. Second we tune the parameter in the Lorenz system to obtain double stable eigenvalues at the origin to showcase our method in this context.

In the final example the validated computation of stable and unstable manifolds is used as ingredient for the rigorous computation of connecting orbits. In particular we consider also the case of regular resonant eigenvalues. Specifically we compute connecting orbits in the system

$$\begin{cases} u'' = -\frac{1}{4}\gamma u - \frac{\sqrt{2}}{4}v^2 + \frac{3}{8}u^3 + 3uv^2, \\ v'' = -\gamma v - \frac{\sqrt{2}}{2}uv + 9v^3 + 3u^2 v, \end{cases} \tag{1}$$

which arises as amplitude equations for the pattern formation model

$$\partial_t U = -(1+\Delta)^2 U + \mu U - \beta \left|\nabla U\right|^2 - U^3, \tag{2}$$

as shown in Doelman et al. (2003). Here $U = U(t, x)$, $t \geq 0$ and $x \in \mathbb{R}^2$. Equation (2) arises in the study of the interplay between trivial, hexagonal and roll patterns near the onset of instability of the zero solution. The parameters γ,μγ,μ and ββ are related via $\gamma = \frac{\mu}{\beta^2}$. See Doelman et al. (2003) and van den Berg et al. (2015) for further details for the relation between (1) and (2).

Following the approach of van den Berg et al. (2015) we prove the existence of a heteroclinic connection between the hexagon and ground states. The proof is based on rigorous numerics for a boundary value problem, where the validated manifolds are used to formulate the boundary conditions. We settle a case left open in van den Berg et al. (2015) due to the presence of resonances in the stable eigenvalues at the origin. In Fig. 1 we depict the verified connecting orbit of (1) as well as the corresponding stationary transition layer between hexagonal spots and the uniform state of (2).

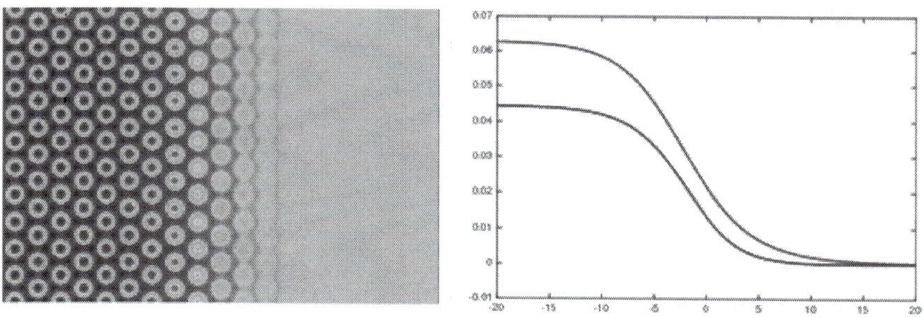

Figure 1. *Left* co-existing hexagonal and trivial patterns in (2) for $\beta = 1$ and $\mu = -\frac{2}{135}$. *Right* corresponding connecting orbit in (1) with *u* and *v* components in *red* and *blue*, respectively (Color figure online)

We remark that an interesting future extension of this work would be to apply the ideas developed here to the validated computation of local stable/unstable manifolds of periodic orbits. The parameterization method has already been adapted to this context. See Guillamon and Huguet (2009), Huguet and de la Llave (2013) and Castelli et al. (2015) for more details and numerical examples. Using the techniques of the present work it should be possible to validate these computations even in the presence of a resonance between the Floquet multipliers. This will be the subject of a future study.

Finally we remark that the references mentioned in this introductory discussion are far from exhaustive, and a comprehensive overview of the literature is beyond the scope of the present work. In recent years a number of authors have developed numerical validation procedures which provide mathematically rigorous a posteriori error bounds on approximations of invariant manifolds associated to various kinds of invariant sets. We refer the interested reader to CAPD (2015), Capinksi and Simo (2012), Johnson and Tucker (2011), Wittig et al. (2010) and Wittig (2011) for fuller discussion of methods other than those presented here.

The outline of the paper is as follows. Section 2 is dedicated to the description of the general setup of our approach. In Sect. 3 we give more details on how we derive the zero finding problem. In Sect. 4 we transform it to an equivalent local fixed point problem to be solved by a parametrized Newton–Kantorovich-type argument. In Sect. 5 we illustrate the performance of our method with the three examples described above. The code implementing these examples can be found at the webpage Code page (2015).

2 SETUP

We consider the validated computation of a parametrization of the local stable manifold of a hyperbolic fixed point p∈Rn of a dynamical system induced by a nonlinear ODE

$$u' = g(u) \qquad g : \mathbb{R}^n \to \mathbb{R}^n \tag{3}$$

using the parametrization method developed in Cabré et al. (2003a, b, 2005). Local unstable manifolds can be obtained by replacing g with −g−g. We assume that g is analytic, allowing us to look for parametrizations in the analytic category. In particular we assume that g is locally (near p) analytically extendable to the complex plane. As a consequence the coefficients in the power series expansion of the parametrization decay geometrically, at an a priori unknown rate. We come back to the role of this decay rate later, see Remark 2.1 and in particular Sects. 5.1.1 and 5.1.2.

2.1 **The Invariance Equation**

We denote by $\lambda_1, \dots, \lambda_d$ the eigenvalues with negative real part of the Jacobian *Dg*(*p*) at the fixed point *p*. To fix notation, let there be *s* pairs of complex conjugate $\lambda_1, \lambda_2, \dots, \lambda_{2s-1}, \lambda_{2s}$ with negative real part and d−2sd−2s real negative eigenvalues $\lambda_{2s+1}, \dots, \lambda_d.$. The corresponding (generalized) eigenvectors are denoted by ξ1,...,ξd. We do not assume that the algebraic multiplicity of the eigenvalues is one. For simplicity, in this paper we assume *p*, λi and ξi to be a priori determined analytically. However, it is straightforward to append equations for the equilibrium, as well as for the linearization around it, to the computational part of the analysis.

We call the set of stable eigenvalues non-resonant if for every i=1,...,di=1,...,d

$$k \cdot \lambda \neq \lambda_i \tag{4}$$

for all $k = (k_1, \dots, k_d) \in \mathbb{N}^d \text{ with } k \neq e_i$ (the *i*-th unit vector). We shall use the notation

$$|k| \stackrel{\text{def}}{=} \sum_{i=1}^{d} k_i.$$

As a new feature of the present work in comparison with the analysis in van den Berg et al. (2011), Mireles-James and Mischaikow (2013) and Mireles-James (2015) we are able to incorporate resonant cases directly in our novel framework for solving and validating the parametrization of the (un)stable manifold. In particular, we focus on the two types of *co-dimension one resonances*, namely a single *regular* resonance and an algebraically double, geometrically simple eigenvalue. We have a *regular* resonant case when

$$\tilde{k} \cdot \lambda = \lambda_{\tilde{i}} \tag{5}$$

for some $\tilde{i} \in \{1,\ldots,d\}$ and a $\tilde{k} \in \mathbb{N}^d$ with $|\tilde{k}| \geq 2$. We work out in detail the cases where a regular resonance occurs as the only resonance, and where a double real eigenvalue occurs as the only resonance. In both cases, in the co-dimension one situation, the resonant eigenvalue is *real* valued. Higher co-dimension resonances, in particular multiple or simultaneous resonances (combinations of regular resonances and/or eigenvalues with higher multiplicity), can be dealt with analogously in our general framework.

In the parametrization method one looks for a map *P* that conjugates the flow of (3) on the stable manifold to a *d*-dimensional flow θ′=h(θ) for a suitably simple choice of *h*. Specifically, let P:Cd⊃Bν→Cn be analytic on the complex polydisc

$$\mathbb{B}_\nu \stackrel{\text{def}}{=} \left\{z \in \mathbb{C}^d : |z_i| \leq \nu_i\right\}, \tag{6}$$

where ν=(ν1,...,νd) with νi>0. We commonly refer to its domain as the parameter space. In particular, the map *P* possesses a *d*-variate series expansion

$$P(\theta) = \sum_{|k|=0}^{\infty} a_k \theta^k \tag{7}$$

with ak=(ak1,...,akn)∈Cn for |k|≥0. We use the usual multi-index notation, where for k=(k1,...,kd) we set dθk=θ1k1...θdkd. The invariance equation for *P* is given by

$$g(P(\theta)) = DP(\theta)h(\theta), \qquad \theta \in \mathbb{B}_\nu. \tag{8}$$

Additionally and without loss of generality, we prescribe the linear constraints

$$a_0 = p$$
$$a_{e_i} = \xi_i \qquad \text{for } i = 1,\ldots,d.$$

Remark 2.1

Note that the choice of the eigenvectors ξi (i=1,...,d) is not unique.

In Lemmas 2.2, 2.3 and 2.5 we analyze the relation between the domain radius $\nu\nu$ in (8) and the lengths $\|\xi i\|$. It will turn out that from a numerical perspective we can either vary ν or $\|\xi i\|$. See Remarks 5.1, 5.2and Breden et al. (2015), Falcolini and de la Llave (1992) and Cabré et al. (2003a) for a more thorough discussion of this topic.

Concerning the choice for the general polynomial normal form of $h(\theta)$ in (8), note that in the non-resonant and double eigenvalue case it is sufficient to use information on the eigenvalues $\lambda 1,\ldots\lambda d$ to do so. In the resonant case we choose a polynomial ansatz informed by spectral information but solve for the corresponding coeffient [τ in (27)]. In all cases we shall choose it such that the origin is a (globally) attracting sink in parameter space. We will also see that we can find a subset $B\nu\hat{}\subset B\nu$ such that the orbits under the flow $\theta'=h(\theta)$ of initial data in $B\nu\hat{}$ do not leave $B\nu B\nu$. We establish the explicit relation between $\nu\nu$and $\nu\hat{}$ in the three cases under consideration in Lemma 2.7. By the conjugation property of *P* and (9a), a suitable real-valued restriction (see below) of the image of$B\nu\hat{}$ under *P* thus gives us a parametrization of the local stable manifold $Ws(p)$. We explain the conjugacy of the flows in Sect. 2.5.

Let us describe how we can define a real-valued parametrization of the stable manifold starting with a complex valued solution of (8) fulfilling (9). Recall that $\lambda = (\lambda_1, \ldots, \lambda_{2s}, \lambda_{2s+1} \ldots \lambda_d)$ consists of *s* pairs of complex conjugate eigenvalues and $d-2sd-2s$ real eigenvalues. We introduce the involutory permutation matrix $\sigma s\sigma s$ of the form

$$\sigma_s \stackrel{\text{def}}{=} \begin{pmatrix} \sigma & 0 & 0 & 0 \\ 0 & \ddots & 0 & 0 \\ 0 & 0 & \sigma & 0 \\ 0 & 0 & 0 & I_{d-2s} \end{pmatrix} \in \mathbb{R}^{d,d} \qquad \text{with} \qquad \sigma = \begin{pmatrix} 0 & 1 \\ 1 & 0 \end{pmatrix}, \tag{10}$$

and $Id-2s$ denoting the $d-2s$-dimensional identity matrix. In essence, this involution plays the role of a symmetry. We introduce an involution on *d*-tuples (including but not limited to Cd, Rd and Nd by

$$(v^*) \stackrel{\text{def}}{=} \sigma_s \overline{v}, \tag{11}$$

where $\overline{v}$ denotes complex conjugation. In particular, we have the involution k↦k∗ on Nd. We have ordered the eigenvalues so that λ∗=λ, and normalized the (generalized) eigenvectors so that ξ∗=ξ. Furthermore, for any variables q=(qk)k∈Nd that allow complex conjugation, we denote (again an involution)

$$(q^*)_k \stackrel{\text{def}}{=} \overline{q}_{k^*}. \tag{12}$$

Next, consider the set

$$\mathbb{B}_\nu^{\text{sym}} \stackrel{\text{def}}{=} \{\theta \in \mathbb{B}_\nu : \theta^* = \theta\}, \tag{13}$$

for any compatible choice of ν∈R+d, i.e., ν∗=ν. The set BsymνBνsym is *d*-dimensional when interpreted as a real linear space (intersected with the ball BνBν), and in the particular case of all eigenvalues being real this boils down to Bνsym={θ∈Rd,|θi|≤νi}. Moreover, we will find, see Sect. 4, that the coefficients *a* of *P*have the symmetry property

$$a^* = a. \tag{14}$$

Under condition (14) we obtain the following lemma.

Lemma 2.1

Assume (14) to be fulfilled. The map *P* is *real*-valued on the invariant subspace BsymνBνsym.

Proof 2.1

Using the properties of the involution (11) in the second equality, together with θ∗=θθ∗=θ (on BsymνBνsym) and a∗=aa∗=a(assumed) in the third, we compute

$$\overline{\sum_{k\in\mathbb{N}^d} a_k\theta^k} = \sum_{k\in\mathbb{N}^d} \bar{a}_k\bar{\theta}^k = \sum_{k\in\mathbb{N}^d} a^*_{k^*}\theta^{*k^*} = \sum_{k\in\mathbb{N}^d} a_{k^*}\theta^{k^*} = \sum_{k\in\mathbb{N}^d} a_k\theta^k,$$

where the last equality uses the invariance of the summation domain under the involution $*$.

Together with Lemma 2.6 which explains the conjugation property of P in more detail, this establishes that P restricted to Bsymv^Bv^sym parametrizes the real local stable manifold of p (see Lemma 2.7 for the relation between νvand ν^).

Our goal is to compute a numerical approximation of P together with rigorous bounds on the approximation error and its range of validity BνBν by using the method presented in Day et al. (2007). This amounts to first formulating an equivalent zero finding problem on an appropriate Banach space. Second, using an approximate zero, we define a Newton-like fixed point operator T. We establish contractivity of T on a ball around the approximate zero by deriving bounds on the residual, as well as bounds on the derivative that depend polynomially on the radius of the ball. These bounds are used to define so-called radii polynomials as ingredients for a finite set of inequalities encoding the prerequisites for the Banach fixed point theorem. We stress that in this way the radius of the ball on which we obtain contractivity is a variable for which we solve. This is an essential difference of the method in Day et al. (2007) compared to classical Newton–Kantorovich-type arguments. Let us assemble the ingredients to define the zero finding problem.

2.2 Non-resonant Eigenvalues

We distinguish two approaches for solving the invariance equation: the *recursive approach* and the *zero finding approach*. The zero finding approach will pave the way to an application of a fixed point argument in the space of power series coefficients. To be concrete, in order to explain the difference between the two approaches, we consider first the case that λ1,...,λd are non-resonant. In this case we choose

$$h(\theta) = \Lambda_s \theta, \tag{15}$$

where ΛsΛs is the diagonal matrix containing the stable eigenvalues on the diagonal. Clearly, the origin is the global attractor for θ′=h(θ). Moreover, h(θ∗).

By substituting the series expansion (7) into the invariance Eq. (8), we derive equations for the series coefficients

This leads to the *homological equations* for all |k|≥2:

$$(Dg(p) - k \cdot \lambda I_n)\, a_k = b_k, \tag{16}$$

where bk only depends on ak^ with |k^|<|k and InIn denotes the *n*-dimensional identity matrix. Note that bkbkvanishes for |k|=1; hence, (17) reduces to the eigenvalue-eigenvector equation, which is solved by (9b).

Using the initial constraints (9) for ak with |k|=0,1 and the fact that λ1,...,λd are non-resonant, (17) can be used to compute akak recursively to any desired order k|≤N). This is what we refer to as the*recursive approach*. The recursive approach shows that there is (a priori) a unique solution of (8) satisfying the constraints (9), although the decay of the sequence is not guaranteed a priori. The validation in van den Berg et al. (2011), Mireles-James and Mischaikow (2013) and Mireles-James (2015) relies on analysis in function spaces of so-called *N*-tails.

Equation (17) is derived by writing g(P(θ)) as a power series expansion in θ and matching like powers in the left- and right-hand sides of (8). In particular,

$$g(P(\theta)) = \sum_{|k|=0}^{\infty} c_k \theta^k \qquad \text{with} \qquad c_k = c_k(a) = c_k\big(\{a_{\hat{k}}, \hat{k} \preceq k\}\big), \tag{17}$$

where the $\hat{k} \preceq k$ means $\hat{k}_i \leq k_i$ for $i = 1, \ldots, d$. For later use, we also introduce the notation $\hat{k} \prec k$ for those $\hat{k} \preceq k$ with $\hat{k} \neq k$. Substituting (18) into the invariance Eq. (8) leads to the equations

$$c_k = (k \cdot \lambda)\, a_k. \tag{18}$$

By observing that bkbk is defined by the splitting

$$c_k = Dg(p)a_k + b_k, \tag{19}$$

where bk depends only on $\{a_k = (a_k^1, \ldots, a_k^n)\}_{k \in \mathbb{N}^d},$ one derives (17). In contrast, in the *zero finding approach* we omit the splitting (20) and instead interpret (19) as a zero finding problem

$$(k \cdot \lambda)\, a_k - c_k = 0 \qquad \text{for all } |k| \geq 2 \tag{20}$$

on a space of geometrically decaying sequences $\{a_k = (a_k^1, \ldots, a_k^n)\}_{k \in \mathbb{N}^d},$ see Sect. 3 and also Beyn and Kless (1998) for a related approach.

The recursive approach shows that the initial constraints (9) for akak with |k|=0,1 determine akak for k|≥2uniquely.
While a0=pa0=p is uniquely fixed by the problem at hand, we enjoy the freedom to scale akak for |k|=1 as those are given by the eigenvectors of *Dg*(*p*). The following lemma illuminates how such a scaling influences akak for |k|≥2. Let μ∈Cd be given such that μ∗=μ We define the scaling (μkak)k∈Nd. In this way scaling preserves the symmetry (14). Moreover, for any analytic nonlinearity *g*, it follows from the power series representation (18) that

$$c_k(\mu a) = \mu^k c_k(a). \tag{21}$$

We have the following invariance of the conjugation map under rescaling by μμ.

Lemma 2.2

Let $a = (a_k)_{k \in \mathbb{N}}$ *fulfill* (9) together with (21) for all k|≥2. Then μaμa also solves (21) for all |k|≥2, whereas $(\mu a)_{e_i} = \mu_i \xi_i$ *for* $i = 1, \ldots, d.$

This follows from (22) and (21). For more detailed discussion see Breden et al. (2015) and also Falcolini and de la Llave (1992) and Cabré et al. (2003a).

We note that the above lemma is equivalent to the observation that the scaling $\theta_i \to \mu_i \theta_i$ leaves the flow in parameter space invariant, and hence by the conjugacy property, $\mu_\alpha a_\mu$ solves the homological equations whenever *a* does. In Remark 5.1 we come back to the practical implications of this scaling invariance.

2.3 **Double Eigenvalues**

We now consider a single repeated (real) eigenvalue with geometric multiplicity one (and no other resonances). Assume without loss of generality that λ_{2s+2} with the rest of the eigenvalues being distinct and non-resonant. We choose ξ_{2s+1} to be a (real valued) eigenvector for the double eigenvalue, and ξ_{2s+2} a (real valued) generalized eigenvector such that

$$[Dg(p) - \lambda_{2s+1} I_n]\xi_{2s+2} = \xi_{2s+1}. \qquad (23)$$

Furthermore, we choose the flow in parameter space to be (cf. Jordan normal form)

$$h(\theta) = \Lambda_s \theta + \theta_{2s+2} e_{2s+1}, \qquad (24)$$

so that it is compatible with (23) through (8). Note that in the context of parameter-dependent families of vector fields this approach can be numerically stabilized by using versal transformations, see for example Chow et al. (1994). In the current work no numerical instabilities occur however, as we do not consider continuation problems (although this is the subject of ongoing research).

We observe, again, that the origin is the global attractor for $\theta' = h(\theta)$, and that $h(\theta^*) = h(\theta)^*$. The corresponding version of (19) is

$$c_k = (k \cdot \lambda)\, a_k + (k_{2s+1} + 1) a_{k+e_{2s+1}-e_{2s+2}}, \qquad (25)$$

for all $|k| \geq 2$ with $k_{2s+2} \geq 1$ We note that the additional term $(k_{2s+1}+1) a_{k+e_{2s+1}-e_{2s+2}}$ occurs at the same "level of recursion" as the term akak, i.e., |k||k+e2s+1−e2s+2|=|k|. While this necessitates caution in the recursive computation of ak (which can still be done with the correct ordering), it does not introduce difficulties for our interpretation of (25) as a zero finding problem. Concerning the influence of scaling akak with μ∈ (μ∗=μ) we obtain the following.

Lemma 2.3

Let $a = (a_k)_{k \in \mathbb{N}}$ *fulfill* (9) *together with* (24) *for all* $|k| \geq 2$. *Let* $\mu \in \mathbb{C}^d$ be such that $\mu^* = \mu$ *and* $\mu_{2s+1} = \mu_{2s+2}$. Then μaμa also solves (25) for all k|≥2, whereas (μa)ei=μiξi for i=1,...,di=1,...,d.

Proof 2.2

The proof follows from (25) by noticing that μk+e2s+1−e2s+2=μk.

If one chooses a rescaling with μ2s+1≠μ2s+2, the normal form (24) needs to be adapted accordingly.

2.4 **Regular Resonant Eigenvalues**

As discussed before, the two co-dimension one resonances are the (real) double eigenvalue (Sect. 2.3) and the *regular* resonant eigenvalue (the other co-dimension one resonance) given by

$$\tilde{k} \cdot \lambda = \lambda_{\tilde{\imath}} \tag{26}$$

for some ı~>2s (i.e., λı~∈Rλı~∈R) and a k~∈Nd with k~|≥2 and k~∗=k~. Here we assume that no other resonances occur.

It follows from the discussion in Sect. 2.2 that for the choice h(θ)=Λsθ the mathematically equivalent recursive and the zero finding approaches are bound to fail. More precisely, assuming (26) to be fulfilled, it is apparent from (17) that the equation for ak~ is in general not solvable. To resolve this, we modify *h* to the nonlinear "normal form"

$$c_k = (k \cdot \lambda)\, a_k + \tau(k_{\bar{\imath}} + 1)a_{k+e_{\bar{\imath}}-\tilde{k}} \cdot \tag{27}$$

where τ∈R is to be determined later. This alters (19) for k~ (note that k~ı~=0) to

$$c_k = (k \cdot \lambda)\, a_k + \tau(k_{\bar{\imath}} + 1)a_{k+e_{\bar{\imath}}-\tilde{k}} \cdot \tag{28}$$

Note that k~|≥2 and thus k|+1−|k~|<|k|. This implies that (28) is amenable to recursive solving for k≥ as in the non-resonant case. Obviously, for k=k~ Eq. (28) can be solved only if τ satisfies a solvability condition, which will be discussed below. We note that the choice (27) for *h*represents the simplest effective one. It facilitates an application of bordered matrix techniques (Govaerts2000) to solve (28) in an efficient way while also providing a means to obtain uniqueness of ak~, see below. The defining properties of the regular co-dimension one resonance imply that k~∗=k~ and eı~∗=eı~. Hence, for any τ∈R we infer that h(θ)∗h(θ∗)=h(θ)∗. However, since ττ is a priori unknown, one needs to derive that ττ is real indeed, see Eq. (31).

We now turn our attention to solving (28) for k=k~. As we will see below, there is a unique ττ making (28) for k=k~ksolvable for ak~, and by appending a suitable additional constraint we can make this solution unique. Using that k~ı~=0, we rephrase (28) for k=k~ as

$$\underbrace{(Dg(p) - \lambda_{\bar{\imath}} I_n)}_{A}\, a_{\tilde{k}} + \tau a_{e_{\bar{\imath}}} = b_{\tilde{k}}, \tag{29}$$

with bk~ depending only on ak^ with k^<k~. Note that by construction aeı~∈ker ⃞(A), as it is an eigenvector of *Dg*(*p*) corresponding to the eigenvalue λı~. The following lemma shows that appending an additional constraint on ak~ brings about unique solvability of (29) for the pair (τ,ak~)

Lemma 2.4

Let A∈Rn,n have an algebraically simple eigenvalue 0 with eigenvector ξ. Let ζ∈Rn, ζ≠0 be such that ATζ=0. Then ζTξ≠0 and the matrix

$$\begin{pmatrix} \xi & A \\ 0 & \zeta^T \end{pmatrix} \in \mathbb{R}^{n+1,n+1}$$

is non-singular.

Proof 2.3

The proof of this lemma can, for example, be found in Kuznetsov (2004), p. 174.

We choose a real vector ζ∈ker(AT). From Lemma 2.4 we see that there is a unique pair (τ,ak~) satisfying

$$\begin{pmatrix} a_{e_{\tilde{\imath}}} & A \\ 0 & \zeta^T \end{pmatrix} \begin{pmatrix} \tau \\ a_{\tilde{k}} \end{pmatrix} = \begin{pmatrix} b_{\tilde{k}} \\ 0 \end{pmatrix}.$$

This pair (τ,ak~) solves (29), and in addition

$$\zeta^T a_{\tilde{k}} = 0. \tag{30}$$

Moreover, one derives that

$$\tau = \frac{\zeta^T b_{\tilde{k}}}{\zeta^T a_{e_{\tilde{\imath}}}}, \tag{31}$$

showing that the value of τ is independent of the choice of ζ. Furthermore, since bk is real for real *a* by its definition (20), τ is real whenever *a* is (ζ can be chosen to be real valued since ~λı~ is real). In this case we also obtain a scaling result.

Lemma 2.5

Let (ak)k∈N fulfill (9) together with (28) for all |k|≥2 for the unique τ determined by (31). Let μ∈Cdbe such that μ∗=μ. Then μaμa solves (28) for all k|≥2 with τ replaced by τμ=μk~−eı~τ, whereas (μa)ei=μiξi for i=1,…,di=1,…,d.

Proof 2.4

The proof follows from (28) by using $\mu^{\tilde{k}-e_{\tilde{\imath}}}\tau(\mu a)_{k-\tilde{k}+e_{\tilde{\imath}}} = \mu^k \tau a_{k-\tilde{k}+e_{\tilde{\imath}}}$.

Note that the scaling of τμτμ is easily understood by combining (22) with (31). Finally, to a large extent the double eigenvalue case in Sect. 2.3 may be interpreted as a regular resonance with $\tilde{\imath} = 2s+1$ and $\tilde{k} = e_{2s+2}$ and $\tau = 1$ known a priori (fixed by choosing the standard Jordan normal form).

2.5 **Explicit Dynamics in Parameter Space**

Let us explain how the invariance Eq. (8) encodes the conjugation of the flows of $u' = g(u)$ and $\theta' = h(\theta)$ that we denote by $\Phi(t,u)$ and $\Psi(t,\theta)$ for concreteness. In particular we explain when a restriction to a smaller ball Bν^Bν^ is in order. The following lemma contains the key observation and makes the role of *P* as a conjugation of flows precise.

Lemma 2.6

Assume $g(P(\theta)) = DP(\theta)h(\theta)$ *for* $\theta \in \mathbb{B}_\nu$ *and let* $\theta \in \mathbb{B}_\nu$ and letθ∈Bν be chosen such that $\Psi(t,\theta) \in \mathbb{B}_\nu$ *for all* $t \geq 0$. *Then* $u(t) \stackrel{\text{def}}{=} P(\Psi(t,\theta))$ *solves* $u' = g(u), u(0) = P(\theta)$. Furthermore $\Phi(t, P(\theta)) = P(\Psi(t,\theta)))$ *and* $\lim_{t\to\infty} u(t) = p$.

Proof 2.5

Let θ∈Bν be fixed. Recall that $\frac{d}{dt}\Psi(t,\theta) = h(\Psi(t,\theta))$. We compute

$$\frac{d}{dt}P(\Psi(t,\theta)) = DP(\Psi(t,\theta))h(\Psi(t,\theta)) = g(P(\Psi(t,\theta))), \quad (32)$$

where the last equality follows from (8) by the fact that $\Psi(t,\theta) \in \mathbb{B}_\nu$ *for* $t \geq 0$. By definition $u(0) = P(\Psi(0,\theta)) = P(\theta)$

and thus by uniqueness of the solution to the initial value problem $u' = g(u), u(0) = P(\theta)$ *we get* $g(P(\Psi(t,\theta))) = \frac{d}{dt}\Phi(t, P(\theta))$. we get $g(P(\Psi(t,\theta))) = \frac{d}{dt}\Phi(t, P(\theta))$. Together with (32), this yields $\Phi(t, P(\theta)) = P(\Psi(t,\theta)))$ *for* $t \geq 0$. Since 0 is the global attractor for the flow Ψ, we conclude that $\lim_{t\to\infty} u(t) = \lim_{t\to\infty} P(\Psi(t,\theta)) = P(0) = p$.

In resonant cases we may need to restrict θ to a subset of Bν to ensure that P(θ)∈Ws(p). For example, one may choose a smaller ball Bν^⊂Bν to ensure $\theta \in \mathbb{B}_{\hat{\nu}}$ and $t \geq 0$. To be able to formulate a criterion for all three cases simultaneously, we introduce

$$\ell_0(y_1, y_2) \stackrel{\text{def}}{=} \max\left\{|y_1|, |y_2|e^{\min\{0,-1+y_1/y_2\}}\right\}, \tag{33}$$

and

$$\ell(\hat{\nu}) \stackrel{\text{def}}{=} \begin{cases} \hat{\nu} & \text{for the non-resonant case,} \\ \hat{\nu} + \left[\ell_0\left(\hat{\nu}_{2s+1}, \frac{\hat{\nu}_{2s+2}}{|\lambda_{2s+1}|}\right) - \hat{\nu}_{2s+1}\right] e_{2s+1} & \text{for the double eigenvalue case,} \\ \hat{\nu} + \left[\ell_0\left(\hat{\nu}_{\bar{i}}, \frac{|\tau|\hat{\nu}^{\bar{k}}}{|\lambda_{\bar{i}}|}\right) - \hat{\nu}_{\bar{i}}\right] e_{\bar{i}} & \text{for the regular resonant case.} \end{cases} \tag{34}$$

The following lemma establishes both a uniform and a pointwise criterion. The former establishes a parametrization of Wlocs(p), whereas the latter is convenient when analyzing a specific orbit.

Lemma 2.7

Let $g(P(\theta)) = DP(\theta)h(\theta)$ *for* $\theta \in \mathbb{B}_{\nu}$. Let ℓ0ℓ0 and ℓℓ be as defined in (33) and (34).

1. For all three cases: if $\ell(\hat{\nu}) \preceq \nu$, *then* $P(\mathbb{B}_{\hat{\nu}}) \subset W^s(p)$.
2. For the double eigenvalue case: *if* $\theta \in \mathbb{B}_{\nu}$ *and* $\ell_0\left(\theta_{2s+1}, \frac{\theta_{2s+2}}{|\lambda|}\right) \leq \nu_{2s+1}$,

3. For the regular resonant case: $if\ \theta \in \mathbb{B}_\nu\ and\ \ell_0(\theta_{\tilde{\imath}}, \frac{\tau\theta^k}{|\lambda_{\tilde{\imath}}|}) \leq \nu_{\tilde{\imath}},\ then\ P(\theta) \in W^s(p).$

.

Proof 2.6

For the non-resonant case this is a direct consequence of Lemma 2.6 and the fact that Ψi(t,θ)≤θi for all t≥0. For the regular resonant case, the explicit flow for θ′=h(θ) is given by

$$\Psi(t,\theta) = \exp(\Lambda_s t)\theta + \left(\tau\theta^{\tilde{k}} t e^{\lambda_{\tilde{\imath}} t}\right) e_{\tilde{\imath}}.$$

For i≠ı~ we have |Ψi(t,θ)|≤θi for all t≥0. For the resonant coordinate we infer

$$|\Psi_{\tilde{\imath}}(t,\theta)| = |\theta_{\tilde{\imath}} + \tau\theta^{\tilde{k}} t| e^{\lambda_{\tilde{\imath}} t} \leq \ell_0(\theta_{\tilde{\imath}}, \tfrac{\tau\theta^{\tilde{k}}}{|\lambda_{\tilde{\imath}}|}) \qquad for\ all\ t \geq 0.$$

This proves part (c) and part (a) for the regular resonant case follows from the inequality ℓ0(y1,y2). Finally, the proof for the double eigenvalue case follows by putting ,k~=e2s+2 and τ=1 in the above arguments.

3 THE ZERO FINDING PROBLEM

In this section we derive the zero finding problem on the space of geometrically decaying series coefficients whose solution corresponds to a solution *P* of (8) via (7). The functional analytic setup is close to the one utilized in Hungria et al. (2016) with the main difference lying in the convolution structure.

3.1 Spaces and Norms

As we work with analytical parametrizations *P* of the form (7), we consider the complex sequence spaces

$$W^\nu \stackrel{\text{def}}{=} \{w = (w_k)_{k\in\mathbb{N}^d}, w_k \in \mathbb{C} : \|w\|_\nu \stackrel{\text{def}}{=} \sum_{|k|=0}^{\infty} |w_k|\nu^k < \infty\},$$

(35)

with vER+d. Note that if aj=(akj)kENdEWv for j=1,...,nj=1,...,n, then *P* is well defined in BvBv. If we define for w,w~EWv the convolution operation

$$(w * \tilde{w})_k \stackrel{\text{def}}{=} \sum_{\hat{k}\preceq k} w_{k-\hat{k}}\, \tilde{w}_{\hat{k}}, \tag{36}$$

the Banach space (Wv,||·||v) becomes a Banach algebra with multiplication ∗. In particular,

$$\|w * \tilde{w}\|_\nu \leq \|w\|_\nu \|\tilde{w}\|_\nu.$$

Taking scalar constraints necessary in the resonant case into account we define the space

$$\mathcal{X}^\nu_{l_0} \stackrel{\text{def}}{=} \mathbb{C}^{l_0} \times (W^\nu)^n, \tag{37}$$

where l0l0 denotes the number of extra variables (as well as the number of constraints). In particular l0=0 in the non-resonant case and for double eigenvalues, whereas l0=1 for (single) regular resonances. We have chosen this somewhat convoluted general notation to deal with all cases in one framework. The notation naturally allows incorporation of multiple (simultaneous) resonances by taking l0>1. We denote elements xEXl0v by

$$x = (x_{-1}, \ldots, x_{-l_0}; x^1, \ldots, x^n).$$

To avoid using projection operators we use the somewhat awkward looking, but compact, notation {x−l}l=1l0for the scalar part of *x*. We endow Xl0v with the norm

$$\|x\| \stackrel{\text{def}}{=} \max\left(\max_{l=1,\dots,l_0} |x_{-l}|, \max_{j=1,\dots,n} \|x^j\|_\nu \right).$$

(38)

3.2 **Zero Finding Problem: Non-resonant Case**

We define an operator fnonresfnonres on X0ν whose zeros correspond to analytic solutions *P* of (8) subject to the linear constraints (9). Based on (21), we set

$$f_k^{\text{nonres}}(x) \stackrel{\text{def}}{=} \begin{cases} a_k - p & k = 0 \\ a_k - \xi_i & k = e_i \quad (i = 1, \dots, d) \\ (k \cdot \lambda)\, a_k - c_k(a) & |k| \geq 2, \end{cases}$$

(39)

where x=a and Dg(p)ξi=λiξi. Given a vector field *g* and once ‖ξi‖ is chosen, fnonresfnonres is fixed, and via (7) zeros of (39) correspond to a parametrization of the local stable manifold of *p*.

3.3 Zero Finding Problem: Double Eigenvalue

Based on (25), we define the operator fdoublefdouble on X0ν by

$$f_k^{\text{double}}(x) \stackrel{\text{def}}{=} \begin{cases} a_k - p & k = 0 \\ a_k - \xi_i & k = e_i \quad (i = 1, \dots, d) \\ (k \cdot \lambda)\, a_k - c_k & |k| \geq 2 \wedge k_{2s+2} = 0 \\ (k \cdot \lambda)\, a_k + (k_{2s+1} + 1) a_{k+e_{2s+1}-e_{2s+2}} - c_k(a) & |k| \geq 2 \wedge k_{2s+2} \geq 1, \end{cases} \quad (39)$$

(40)

where x=a, and ξi are the (generalized) eigenvectors, as discussed in Sect. 2.3.

3.4 Zero Finding Problem: Regular Resonant Case

In the regular resonant case, we choose a setup with ττ as an extra unknown and (30) as appended equation. Based on (28) we define the operator fregresfregres on X1ν by

$$f_k^{\mathrm{regres}}(x) \stackrel{\mathrm{def}}{=} \begin{cases} \zeta^T a_{\tilde{k}} & k=-1 \\ a_k - p & k=0 \\ a_k - \xi_i & k=e_i \quad (i=1,\dots,d) \\ (k\cdot\lambda)\, a_k - c_k & |k|\geq 2 \wedge k \not\succeq \tilde{k} \\ (k\cdot\lambda)\, a_k + \tau(k_{\tilde{i}}+1)a_{k+e_{\tilde{i}}-\tilde{k}} - c_k(a) & |k|\geq 2 \wedge k \succeq \tilde{k}, \end{cases} \tag{41}$$

where $x=(\tau;a)$ and $\zeta\in\ker((Dg(p)-\lambda_{\tilde{\imath}} I_n)^T)$ is fixed (and real valued). We recall that we use negative indices to number scalar constraints; hence, we have slightly abused notation here by using $k=-1$ to denote the scalar part of fregresfregres, whereas everywhere else $k\in\mathbb{N}^d$.

4 FIXED POINT OPERATOR AND RADII POLYNOMIALS

We are now equipped with operators fnonresfnonres, fdoublefdouble and fregresfregres given by (39) defined on X^0_ν, (40) defined on X^0_ν, and (41) defined on X^1_ν, respectively. We simply use the notation *f* defined on X if this does not lead to confusion. The zero of *f* corresponds to a parametrization of the stable manifold, a fact we still need to make more precise in due course. We follow the setup of Day et al. (2007) and Hungria et al. (2016) and derive an equivalent fixed point operator, whose contractivity on a ball around an approximate solution we establish using the so-called radii polynomials. Similar approaches have been used in previous works such as Eckmann et al. (1984), Arioli and Koch (2010) and Arioli and Koch (2015) and earlier in Lanford (1982). The structure of the fixed point operator $T:X\to X$ is Newton-like, i.e.,

$$T(x) \stackrel{\mathrm{def}}{=} x - Af(x), \tag{42}$$

where *A*, which is specified below, plays the role of an approximate inverse of $Df(\hat{x})$ with $\hat{x}$ an approximate solution to $f(x)=0$. We extend the symmetry from Sect. 2.1 to $x\in X^{l_0}_\nu$ by setting

$$x^* \stackrel{\mathrm{def}}{=} (\overline{x_{-1}},\dots,\overline{x_{-l_0}};(x^1)^*,\dots,(x^n)^*).$$

We define T, and in particular A, in such a way that T allows a well-defined restriction to the closed symmetric subspace Xsym given by

$$\mathcal{X}^{\mathrm{sym}} = \mathcal{X}_{l_0}^{\nu,\mathrm{sym}} \stackrel{\mathrm{def}}{=} \{x \in \mathcal{X} : x^* = x\} \cong \mathbb{R}^{l_0} \times (W^{\nu,\mathrm{sym}})^n, \tag{43}$$

where

$$W^{\nu,\mathrm{sym}} = \{w \in W^\nu : w^* = w\}. \tag{44}$$

The function f defined in the three cases (39), (40) and (41) respects the symmetry.

Lemma 4.1

We $f(x^*) = f(x)^*$.

Proof 4.1

We start by showing $c(a^*) = c(a)^*$. The vector field $u' = g(u)$ is real; hence $g(\overline{u}) = g(u)$. Considering the definition (18) of $c(a)$, we will exploit the identity $g(\overline{P(a,\theta)}) = \overline{g(P(a,\theta))}$, where we write $P(\theta) = P(a,\theta)$ for clarity. We first observe, by using that the summation domain is invariant under the involution $k \mapsto k^*$, that

$$\overline{P(a,\theta)} = \sum_{k\in\mathbb{N}^d} \overline{a}_k \overline{\theta}^k = \sum_{k\in\mathbb{N}^d} (a^*)_{k^*} (\theta^*)^{k^*} = \sum_{k\in\mathbb{N}^d} (a^*)_k (\theta^*)^k = P(a^*,\theta^*).$$

It follows that

$$g(\overline{P(a,\theta)}) = g(P(a^*,\theta^*)) = \sum c(a^*)_k (\theta^*)^k. \tag{45}$$

On the other hand

$$\overline{g(P(a,\theta))} = \sum_{k\in\mathbb{N}^d} \overline{c(a)_k}\,\overline{\theta}^k = \sum_{k\in\mathbb{N}^d} \overline{c(a)_{k^*}}\,\overline{\theta}^{k^*} = \sum_{k\in\mathbb{N}^d} c(a)^*_k(\theta^*)^k. \tag{46}$$

By combining (45) and (46) we infer that

$$\sum_{k\in\mathbb{N}^d} c(a^*)_k(\theta^*)^k = \sum_{k\in\mathbb{N}^d} c(a)^*_k(\theta^*)^k.$$

It follows from uniqueness of the Taylor coefficients that $c(a^*)=c(a)^*$. Using that $\overline{(k\cdot\lambda)} = k^*\cdot$ $\cdot\lambda$ this proves the assertion for the non-resonant case. Additionally, for the case of double eigenvalue and regular resonant cases we observe that the conditions $k_{2s+2} \geq 1 \; and \; k \succeq k$ are invariant under the involution, since $k^*_i = k_i \; for \; i > 2s+1$ Finally, the symmetry of the scalar part f−1f−1 in (41) follows from $\overline{\zeta} = \zeta$ and $\tilde{k}^* = \tilde{k}.$

We now can make the correspondence between zeros of *f* and parametrizations of local stable manifolds more precise.

Lemma 4.2

If $f(x) = 0 \; for \; x \in \mathcal{X}^{\nu}_{l_0}, \; then \; x^* = x.$ In particular a=(x1,...,xn) defines via (7) a parametrization P:Bν^→Rn of the local stable manifold of *p* provided ν^ and νν satisfy the condition ℓ(ν^)⪯ν in Lemma 2.7a.

Proof 4.2

Assume f(x)=0. Then using Lemma 4.1

$$f(x^*) = f(x)^* = 0^* = 0.$$

By uniqueness of zeros of *f* it follows that x∗=x. By construction of *f* in (39), (40) or (41), *P* defined through *a* via (7) solves (8) under the constraints (9a) and (9b) with *h* specified by (15), (24) or (27), respectively. Hence the statement follows by combining Lemmas 2.1 and 2.6.

To define the linear operator *A* that appears in (42), we start by defining a finite dimensional projection πmπmon X. For m=(m1,...,md), we introduce the index set

$$\mathcal{I}_m \stackrel{\text{def}}{=} \{-1, \ldots, -l_0\} \sqcup \{k \in \mathbb{N}^d : k \preceq m\}.$$

(47)

Note that we require m∗=mm∗=m in the case of *s* complex conjugate eigenvalue pairs, so that k∈Im if and only if k∗∈Imk. Using this notation we set

$$\pi_m x \stackrel{\text{def}}{=} (x_k)_{k \in \mathcal{I}_m}.$$

The range of πm can be identified with CM where M=l0+n∏i=1d(mi+1). The symmetry operation descends to an involution Σs on CM through

$$\Sigma_s \pi_m(\overline{x}) \stackrel{\text{def}}{=} \pi_m(x^*). \tag{48}$$

An element xF in the range of πmπm can be lifted to X through the trivial embedding

$$(\iota x_F)_k \stackrel{\text{def}}{=} \begin{cases} (x_F)_k & k \in \mathcal{I}_m \\ 0 & k \notin \mathcal{I}_m \end{cases} \tag{49}$$

We define the "Galerkin" projection

$$f^m(x_F) \stackrel{\text{def}}{=} \pi_m f(\iota x_F).$$

We now assume an approximate zero $\widehat{x}_F$ of f^m of fmfm has been computed (e.g., using a standard Newton method), and that AmAm is a numerical approximation of the inverse of the Jacobian, i.e., $A_m \approx (Df^m(\widehat{x}_F))^{-1}$. We denote by $A^s_m \stackrel{\text{def}}{=} \frac{1}{2}\left(A_m + \Sigma_s \overline{A_m} \Sigma_s\right)$ its symmetrized version, which has the property

$$A^s_m \Sigma_s = \Sigma_s \overline{A^s_m}. \tag{50}$$

Remark 4.1

Since in practice $\widehat{x}^*_F \approx \widehat{x}_F$, it follows from $f(x^*) = f(x)^*$ *that* $A^s_m \approx A_m$. Consequently, $A^s_m Df^m(\widehat{x}_F) \approx I_M$. Furthermore, replacing AmAm by its symmetrization AsmAms is not strictly necessary, since the symmetry of the fixed point, derived in Lemma 4.5 below, can also be obtained from the a priori uniqueness through Lemma 4.2.

We define the linear operator *A* by

$$(Ax)_k \stackrel{\text{def}}{=} \begin{cases} (A^s_m \pi_m x)_k & k \in \mathcal{I}_m, \\ \frac{1}{\lambda_1 k_1 + \ldots + \lambda_d k_d} x_k & k \notin \mathcal{I}_m. \end{cases} \tag{51}$$

By construction, *A* conserves the symmetry, as expressed by the following lemma.

Lemma 4.3

Let *A* be defined by (51). Then $Ax^* = (Ax)^*$.

Proof 4.3

Since $m^* = m$ *and* $\lambda^* = \lambda$, it follows that $(Ax^*)_k = ((Ax)^*)_k$ *for* $k \notin \mathcal{I}_m$. It remains to establish that $A^s_m x^*_F = (A^s_m x_F)^*$, which follows from (48) and (50).

Lemma 4.4

We have T(x∗) In particular, *T* maps Xsym into itself.

Proof 4.4

This follows directly from Lemmas 4.1 and 4.3.

We aim to show that *T* is contraction on a small ball around an approximate zero $\widehat{x} = \iota\widehat{x}_F \in \mathcal{X}$. It follows from Lemma 4.5 that if x^ is symmetric (x^∗=x^) or almost symmetric, then the unique fixed point of *T* in the ball is a symmetric zero of *f* (provided *A* is injective).

Lemma 4.5

Assume that *T* is a contraction on the ball Bx^(r) with Bx^(r)∩Xsym≠∅. Then *T* has a unique fixed pointx in Bx^(r), and x∼∗=x∼. If, in addition, AsmAms is invertible, then x∼ is a zero of *f*, and hence corresponds a parametrization of the real stable manifold.

Proof 4.5

The first part follows from the Banach fixed point theorem. Since *T* leaves Xsym invariant, *T* is a contraction mapping on Bx^(r)∩Xsym≠∅, hence its fixed point lies in Xsym. If AsmAms is injective, then the fixed point x˜ corresponds to a zero of *f*, and the rest of the proof follows directly from Lemma 4.2.

As explained in Remark 4.1, the assertions of Lemma 4.5 also hold if one replaces AsmAms by its unsymmetrized analogue AmAm, provided it is invertible. We now describe how we show the contractivity of *T* on a suitable ball. Let us introduce the operator A†A† given by

$$(A^{\dagger}x) \stackrel{\text{def}}{=} \begin{cases} (Df^m(\widehat{x}_F)\pi_m x)_k & k \in \mathcal{I}_m \\ (\lambda_1 k_1 + \ldots + \lambda_d k_d)\, a_k & k \notin \mathcal{I}_m, \end{cases} \tag{52}$$

which acts as an approximation to the derivative of *f* at ιx^Fιx^F. Note that in particular AA†≈Id . For later use we note that *DT*(*x*)*y* can be split as

$$DT(x)y = (\mathrm{Id} - ADf(x))y = (\mathrm{Id} - AA^{\dagger})y - A(Df(x)y - A^{\dagger}y). \tag{53}$$

Here Id Id denotes the identity on X. We continue by defining bounds that will be used to prove contractivity of *T*.

Definition 4.1

Let $\widehat{x} = \iota\widehat{x}_F \in \mathcal{X}$ *be given. Let* $Y = (Y_{-1}, \ldots, Y_{-l_0}; Y^1, \ldots, Y^n) \in$ be bounds such that

$$|(T\hat{x} - \hat{x})_{-l}| \leq Y_{-l} \qquad \textit{for } l = 1, \ldots, l_0$$
$$\|(T\hat{x} - \hat{x})^j\|_\nu \leq Y^j \qquad \textit{for } j = 1, \ldots, n.$$

(54a,54b)

Let $Z(r) = (Z_{-1}(r), \ldots, Z_{-l_0}(r); Z^1(r), \ldots, Z^n(r))$ *be* l_0+n polynomials in *r* with nonnegative coefficients, such that for all *v*, *w* with ‖v‖≤1 and ‖w‖≤1 the following bounds hold for all r≥0:

$$|(DT(\hat{x} + rv)rw)_{-l}| \leq Z_{-l}(r) \qquad \textit{for } l = 1, \ldots, l_0$$
$$\|(DT(\hat{x} + rv)rw)^j\|_\nu \leq Z^j(r) \qquad \textit{for } j = 1, \ldots, n.$$

(55 a, 55 b)

We define the *radii polynomials* by

$$p_{-l}(r) \stackrel{\text{def}}{=} Y_{-l} + Z_{-l}(r) - r \qquad \textit{for } l = 1, \ldots, l_0$$
$$p^j(r) \stackrel{\text{def}}{=} Y^j + Z^j(r) - r \qquad \textit{for } j = 1, \ldots, n.$$

(56 a, 56 b)

The crux of this definition is that the bounds *Y* on the residue and *Z* on the derivative of *T* can be constructed explicitly, see the examples in Sect. 5. We note that the inclusion of the scalar *r* in (55) trivially scales the bounds on *DT* by a factor *r*. We include this factor here to keep the notation compatible with earlier papers (going back to Yamamoto 1998). The radius *r* of the ball is not fixed a priori and the l0+nl0+nradii polynomials *p*(*r*) are used in the following parametrized version of the Newton–Kantorovich theorem.

Lemma 4.6

Let r>0 be such that p−l(r)<0 for l=1,...,l0l and pj(r)<0 for j=1,...,n. Then *T* is a contraction on Bx^(r) and there is a unique fixed point x~∈Bx^(r) of *T*.

A proof of this lemma can be found, e.g., in Yamamoto (1998), Day et al. (2007) and Hungria et al. (2016).

Remark 4.2

We emphasize two consequences.

1. The l0+n conditions p(r)<0 reduce the validation of zeros of the operators *f* defined on the infinite dimensional spaces X to a finite set of inequalities that can be checked rigorously using interval arithmetic. Note that the inequality p(r)<0 is to be understood component-wise.

2. We can translate p(r)<0 to a statement about the error of the image of the parametrization in phase space. Denote by Pm(θ)=∑k∈Ima^kθkPm(θ)= the approximate parametrization corresponding to x^x^, and by P(θ)=∑k∈Nda~kθk the exact parametrization corresponding to x~x~. Then for all Bνsym and all j=1,...,nj=1,...,n

$$|P^j(\theta) - P^j_m(\theta)| \leq \sum_{k\in\mathbb{N}^d} |\tilde{a}^j_k - \hat{a}^j_k||\theta^k| \leq \|\tilde{a}^j - \hat{a}^j\|_\nu \leq \|\tilde{x} - \hat{x}\| \leq r. \tag{57}$$

3. Another consequence is that

$$|\tilde{a}^j_k| \leq \frac{r}{\nu^k} \qquad \mathit{for\ all}\ k \notin \mathcal{I}_m\ \mathit{and}\ j = 1, \dots, n. \tag{58}$$

This gives control over the tail coefficients in the exact parametrization. Note that it is this information that is crucial to the method in van den Berg et al. (2011) for deriving a posteriori bounds on the derivative of the truncation error in the parametrization (see van den Berg et al. (2011), Section 5.2, Eqn. (87)). As a consequence, the analysis in van den Berg et al. (2011) is applicable with the radius r obtained from the current approach.

To derive the bounds (54) and (55) good analytical control on the operator f and its derivative Df at x^ is essential. In Sects. 5.1 and 5.2 we illustrate the "mechanics" involved in the derivation of these bounds.

5 APPLICATIONS

In this section we consider three applications to illustrate the performance of our method. We start with the well-known Lorenz equations and compute a 2D local stable manifold at the classical parameter values yielding non-resonant eigenvalues. Subsequently, we consider non-standard parameter values in the Lorenz system to investigate the specific issues in the double eigenvalue case. Finally we consider the case of regular resonant eigenvalues in a system of ODEs originating from a pattern formation model. We compute (un)stable manifolds that serve as ingredient for a connecting orbit computation in this system.

5.1 Local Manifolds in the Lorenz System

We consider the Lorenz differential equations (Lorenz 1963)

$$\dot{u} = g(u) = \begin{pmatrix} \sigma(u_2 - u_1) \\ \rho u_1 - u_2 - u_1 u_3 \\ u_1 u_2 - \beta u_3 \end{pmatrix}. \tag{59}$$

In Sect. 5.1.1 we choose the classical parameter values β=83 and σ=10 and ρ=28 and describe the validation process including the derivation of the necessary bounds for the 2D local stable manifold of the origin (local Lorenz manifold) in some detail. In particular, we investigate how to choose the various computational parameters involved in the validation process taking different objectives into account. In Sect. 5.1.2 we use the fact that we have explicit formulas for the stable eigenvalues at the origin to tune the parameters β,ρ,σ such that there are double eigenvalues at the origin. We explain the validation analysis in this context, which serves as preparation for the regular resonant case.

5.1.1 Detailed Analysis for the Local Lorenz Manifold: Non-resonant Eigenvalues

The dimension of the manifold is $d=2$, and the phase space dimension is $n=3$. Denote the stable eigenvalues by $\lambda_{1,2}$ and the corresponding eigenvectors by $\xi_{1,2}$. As a matter of fact $\lambda_{1,2}\in\mathbb{R}$. Hence, we search for a map $P:\mathbb{R}^2\supset B_\nu\to\mathbb{R}^3$ together with $\nu=(\nu_1,\nu_2)$ fulfilling (8), with *g* given by the Lorenz vector field (59), and the linear constraints (9). As we are in the non-resonant case we can choose $\nu=\hat{\nu}$ as discussed in Sect. 2.5. We make the power series ansatz (7) with $a_k=(a^1_{k_1k_2},a^2_{k_1k_2},a^3_{k_1k_2})\in\mathbb{R}^3$. Note that the coefficients $c_k=(c^1_{k_1k_2},c^2_{k_1k_2},c^3_{k_1k_2})$ in the expansion (18) are

$$c_{k_1k_2} = \begin{pmatrix} \sigma\left(a^2_{k_1k_2} - a^1_{k_1k_2}\right) \\ \rho a^1_{k_1k_2} - a^2_{k_1k_2} - (a^1 * a^3)_{k_1k_2} \\ (a^1 * a^2)_{k_1k_2} - \beta a^3_{k_1k_2} \end{pmatrix} \tag{60}$$

with $*$ defined in (36). Together with $p=0\in\mathbb{R}^3$ this completes the ingredients for fnonresfnonres defined in (39).

Recalling the definition of the radii polynomials in (56b), we notice that a necessary condition for finding a radius *r* fulfilling $p_i(r)<0$ for $i=1,2,3$ is that the components of Z_1 derived in (65) be smaller than one. The parameters that are under our direct control are $m=(m_1,m_2)$ and $\nu=(\nu_1,\nu_2)$ or $\|\xi_{1,2}\|$ respectively. Note that varying $\nu_{1,2}$v1,2 is in the following precise sense *equivalent* to varying $\xi_{1,2}$, which implies that we may as well fix either νν or ξ.

Remark 5.1

From the scaling operation analyzed in Lemma 2.2, we notice that

$$\|\mu a^j\|_\nu = \|a^j\|_{\mu\nu}, \qquad j = 1,2,3.$$

This implies that we should *either* vary the decay rates (domain radius) νν *or* the (eigenvector) scalings μμ to affect ‖a^j‖ν, j=1,2,3j=1,2,3 in Z1. Below we fix ν=(1,1), as this is numerically the most stable choice.

Let us describe the dependence of the coefficients of pi(r)pi(r) on these computational parameters in more detail by deriving explicit formulas for the radii polynomials.

Derivation of the bounds: details To define the radii polynomials specified in (56b) we first need to compute the bounds Yj and Zj (j=1,2,3) defined in (54) and (55). Assume we have already calculated an approximate zero x^=(a^1,a^2,a^3)=ιx^F where ιι is defined in (49), and set Am≈(Dfm(x^F))−1 We start by noting that (f(x^))k=0(f(x^))k=0 for k∉I2mk∉I2m, since *g*(*u*) is quadratic. Using this and the fact that T(x^)−x^=Af(x^) we set yjk=(|(Amfm(x^))jk|) for k∈Im, yjk=1k1|λ1|+k2|λ2|((|f2m(x^))jk|) for for k∉I2mk∉I2m. The bounds Yj (j=1,2,3) are then obtained by computing the finite sums ‖yj‖ν‖yj‖ν.

To derive the bounds Zj(r) defined in (55) we use the splitting (53). Let v,w∈B1(0). We start by deriving an expansion

$$(Df(\hat{x}+rv)rw - A^{\dagger}rw)_k = z_{k,1}r + z_{k,2}r^2 \qquad \text{with } z_{k,1}, z_{k,2} \in \mathbb{R}^3. \tag{61}$$

The explicit expressions for *z* are listed in Table 1.

Our next goal is to compute a bound

Table 1. Expansion coefficients for $(Df(\bar{x}+rv)rw - A^{\dagger}rw)_k$ in (59).

	$k \in \mathcal{I}_m$	$k \notin \mathcal{I}_m$
$z_{k,1}$	$\begin{pmatrix} 0 \\ 0 \\ 0 \end{pmatrix}$	$\begin{pmatrix} \sigma(w_k^2 - w_k^1) \\ \rho w_k^1 - w_k^2 - ((\hat{a}^1 * w^3)_k + (\hat{a}^3 * w^1)_k) \\ (\hat{a}^1 * w^2)_k + (\hat{a}^2 * w^1)_k - \beta w_k^3 \end{pmatrix}$
$z_{k,2}$	$\begin{pmatrix} 0 \\ -((w^1 * v^3)_k + (w^3 * v^1)_k) \\ (w^1 * v^2)_k + (w^2 * v^1)_k \end{pmatrix}$	$\begin{pmatrix} 0 \\ -((w^1 * v^3)_k + (w^3 * v^1)_k) \\ (w^1 * v^2)_k + (w^2 * v^1)_k \end{pmatrix}$

$$|(DT(\hat{x}+rv)rw)_k^j| \le Z_{k,1}^j r + Z_{k,2}^j r^2. \quad (62)$$

The first term to estimate is (Id −AA†)w. We introduce the notation Am†= def Dfm(x^F), and by using |wkj|≤1νk we compute, by a direct computation, ϵk∈R3, k∈Im such that

$$|((\mathrm{Id} - AA^{\dagger})w)_k^j| \le (|\mathrm{Id} - A_m A_m^{\dagger}|\,|w_F|)_k^j \le \epsilon_k^j \qquad \text{for all } k \in \mathcal{I}_m, \quad (63)$$

where absolute values are to be understood component-wise. Furthermore, for the tail terms we have ((Id −AA†)w)k=0∈R3 for k∉Im by construction. To accommodate matrix multiplication we collect all estimates of zk,2 for k∈Im in a vector. Namely, we define the vector χ2∈R3nmχ2∈R3nm with nm=(m1+1) such that

$$|z_{k,2}^j| \le \chi_2^{3(k_2(m_1+1)+k_1)+j} \qquad \text{for } 0 \le k_1 \le m_1, 0 \le k_2 \le m_2, j = 1,2,3. \quad (64)$$

To be explicit, we use the estimate

$$\chi_2^{3(k_2(m_1+1)+k_1)+1} = 0, \quad \chi_2^{3(k_2(m_1+1)+k_1)+2} = \frac{2}{\nu^k}, \quad \chi_2^{3(k_2(m_1+1)+k_1)+3} = \frac{2}{\nu^k}.$$

Then applying the definition of A given in (51), and interpreting AmAm as an 3nm×3nm3nm×3nm matrix, yields the values for Zk,1 and Zk,2 summarized in Table 2.

Table 2 Expansion coefficients for $Z_k(r) = Z_{k,1}r + Z_{k,2}r^2$. All absolute values are to be understood component-wise. The values of ϵ_k are defined in (63)

	$k \in \mathcal{I}_m$	$k \notin \mathcal{I}_m$
$Z_{k,1}$	ϵ_k	$\frac{1}{k_1\lvert\lambda_1\rvert+k_2\lvert\lambda_2\rvert}\begin{pmatrix}\sigma(\lvert w\rvert_k^2+\lvert w\rvert_k^1)\\ \rho\lvert w\rvert_k^1+\lvert w\rvert_k^2+((\lvert\hat{a}\rvert^1*\lvert w\rvert^3)_k+(\lvert\hat{a}\rvert^3*\lvert w\rvert^1)_k)\\ (\lvert\hat{a}\rvert^1*\lvert w\rvert^2)_k+(\lvert\hat{a}\rvert^2*\lvert w\rvert^1)_k+\beta\lvert w\rvert_k^3\end{pmatrix}$
$Z_{k,2}$	$(\lvert A_m\rvert\chi_2)_k$	$\frac{1}{k_1\lvert\lambda_1\rvert+k_2\lvert\lambda_2\rvert}\begin{pmatrix}0\\ ((\lvert w\rvert^1*\lvert v\rvert^3)_k+(\lvert w\rvert^3*\lvert v\rvert^1)_k)\\ (\lvert w\rvert^1*\lvert v\rvert^2)_k+(\lvert w\rvert^2*\lvert v\rvert^1)_k\end{pmatrix}$

Finally, by estimating the finite part ($k \in \mathcal{I}_m$) and the tail part ($k \notin \mathcal{I}_m$) separately, we can compute $Z_j(r) = Z_1^j r + Z_2^j r^2$ by

$$Z_1 = \sum_{k\in\mathcal{I}_m} \epsilon_k \nu^k + \frac{1}{\min\{(m_1+1)\lvert\lambda_1\rvert,(m_2+1)\lvert\lambda_2\rvert\}}\begin{pmatrix}2\lvert\sigma\rvert\\ \lvert\rho\rvert+1+\lVert\hat{a}^1\rVert_\nu+\lVert\hat{a}^3\rVert_\nu\\ \lvert\beta\rvert+\lVert\hat{a}^1\rVert_\nu+\lVert\hat{a}^2\rVert_\nu\end{pmatrix} \tag{65a}$$

$$Z_2 = \sum_{k\in\mathcal{I}_m} (\lvert A_m\rvert\chi_2)_k\nu^k + \frac{1}{\min\{(m_1+1)\lvert\lambda_1\rvert,(m_2+1)\lvert\lambda_2\rvert\}}\begin{pmatrix}0\\2\\2\end{pmatrix}, \tag{65b}$$

where we have used the Banach algebra convolution estimate in W^ν to bound the tail sums. This completes the ingredients for (56b).

The main influence of ν on these estimates is via the terms $\lVert\hat{a}^i\rVert_\nu$ $(i = 1, 2, 3)$. On the other hand these norms are also controlled by the length of the stable eigenvectors $\xi_{1,2}$ appearing in definition (39), and also, albeit weakly, by $m=(m_1,m_2)$. Let us analyze this interplay.

Choice of m and $\lVert\xi_{1,2}\rVert$ The considerations that lead to the settings of the computational parameters depend on the goals in the application at hand. One might, for example, be interested in uniformly maximizing the image in phase space of the parametrization, or one might want to capture a particular point in phase space in the image, i.e., maximize the image in a certain direction. In the following we collect several observations for either task. Let us begin with a fundamental restriction; namely, the components Z_1^j in (65) need to be smaller than one. This can be used as a first feasibility check for the validation.

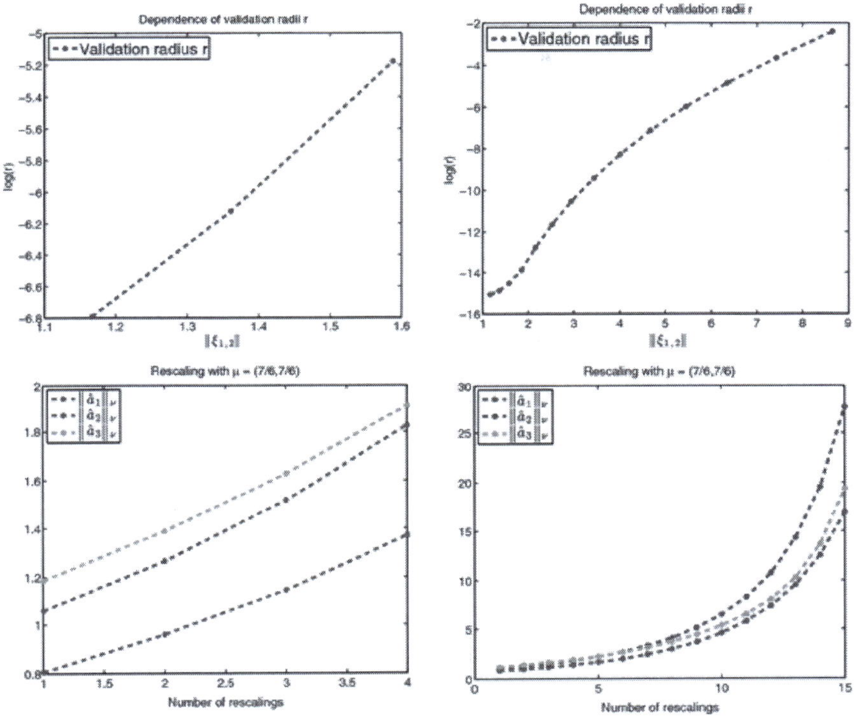

Figure. 2 *Top* change of the validation radii while rescaling the eigenvectors (*left* m=(5,5) *right* m=(15,15) We see that for a larger number of modes we obtain smaller error bounds. Note in addition that the larger the norm of ξ1,2 is the bigger the uniform error bound *r* on BνBν gets. *Bottom* dependence of the norm of the approximate solution on the number of rescalings, hence on the norms ||ξ1,2||. These are an indicator for the size of the image in phase space

This motivates the following procedure (recall that we have fixed ν=(1,1), see Remark 5.1):

1. Choose an order m=(m1,m2)]. Compute a^ with ||ξ1,2||=1.

2. Attempt to check the conditions of Lemma 4.6 using the formulas derived for (54) and (56b).

- In case of failure rescale $\xi1,2=\mu1,2\xi1,2$ and $\hat{}a\hat{}=\mu a\hat{}$ with $0<\mu i<10<\mu i<1$ $(i=1,2)$ and repeat the second step.

- In case of success rescale $\xi1,2$ and $a\hat{}=\mu a\hat{}$ with $\mu i>0$ $(i=1,2)$ chosen according to the maximization objective and repeat the second step until stop at failure.

In this procedure there remain two open choices, namely of $m=(m1,m2)$ and $\mu=(\mu1,\mu2)$. Let us review two options. More thorough discussion of the choice of scalings, including algorithms and implementations, can be found in Breden et al. (2015).

1. $m1=m2$ and $\mu1=\mu2>1$ uniformly maximizing the image

We consider the two cases $m=(5,5)$ and $m=(15,15)$, see Fig. 2. First, for $m=(5,5)$, the validation succeeds with $\|\xi1,2\|=1$ with a radius of $r\approx10-7$. Recall that by Remark 4.2 the validation radius *r* can be seen as an accuracy measure. We thus consider smaller radius *r* as higher accuracy. After conducting 5 rescalings with factors $\mu1=\mu2=76$ to uniformly maximize the image, we fail to validate. The accuracy *r* we obtain after 4 rescalings decreased to $10-510-5$ (for fixed *m* the (uniform estimate on the) accuracy naturally decreases when increasing the domain). Second, for $m=(15,15)m=(15,15)$ we also succeed to validate for $\|\xi1,2\|=1$ but with smaller uniform error bound $r\approx10-15$ We are able to rescale 15 times with the same factors. This increases $\|\xi1,2\|$ to 10.09 and increases the validation radius to $10-210-2$.

2. Fast-slow choice of *m* and $\mu\mu$: maximizing the image in the slow direction

Next we consider the cases $m1\neq m2$ and/or $\mu1\neq\mu2$, see Fig. 3. We recall that $|\lambda1|\approx10|\lambda2||$, hence we refer to $\lambda1/2\lambda1/2$ as the fast/slow eigenvalue. For most orbits the dynamics close to the origin is dominated by the slow direction. This is, for example, of interest when computing connecting orbits that approach the equilibrium along the slow direction. Capturing a large portion of the slow direction can thus be desirable. We choose $m2\geq m1$ and $\mu2>1>\mu1$. First we choose $m1=m2=15$. We succeed to validate for $\|\xi1,2\|=1\|\xi1,2\|=1$ with a radius $r \approx 10-15$. Let $\mu 1=67$ and $\mu 2=76$. We obtain 14 successful rescalings with gradually decreasing accuracy ($r \approx 10-3r \approx 10-3$ after 14 rescalings). If we choose $m1=5$ and $m2=15$ we observe qualitatively different behavior of the validation radii.

Starting with a success at $\|\xi 1,2\|=1$ and radius $r \approx 10-8$ the accuracy increases to $10-11$ in the first 8 rescalings with the factors $\mu 1=67$ and $\mu 2=76$ until the norms of $\|\xi 1,2\|$ "align" with the choice of *m*. Then the accuracy decreases to $10-310-3$ after 18 rescalings.

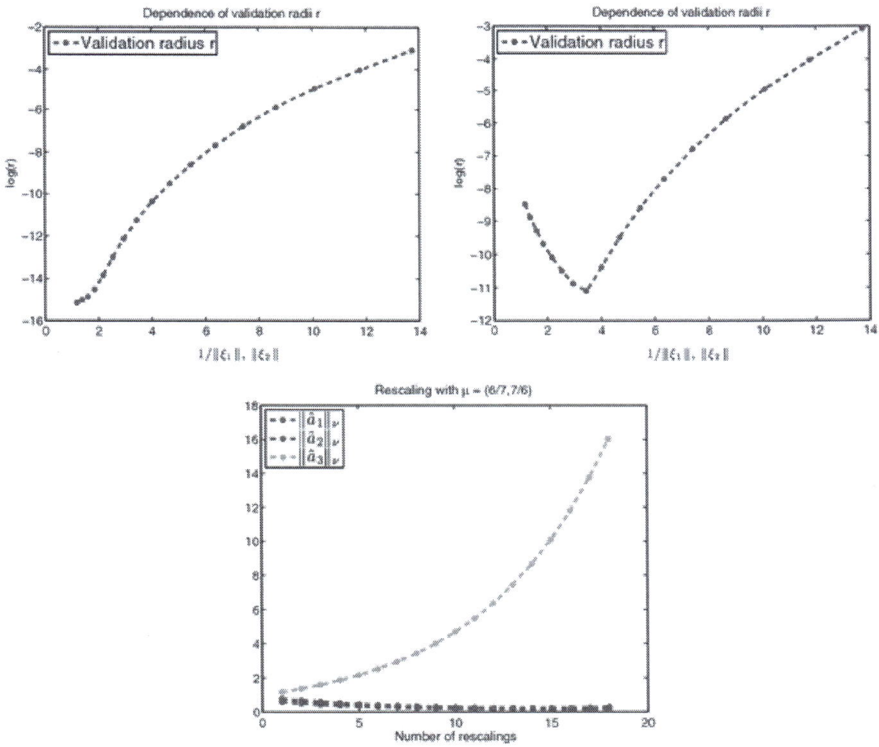

Figure 3. *Top left* change of validation radii while rescaling with factor $\mu=(67,76)$ for the choice $m1=m2=15$. We observe qualitatively similar behavior to the uniform scaling. *Top right* change of validation radii while rescaling with factor $\mu=(67,76)$ for the choice $m1=5,m2=15$. We observe qualitatively different behavior to the uniform scaling. The maximal accuracy is obtained for $\|\xi 2\|\|\xi 1\| \approx 11.8$. *Bottom* dependence of the norm of the approximate solution on the number of rescalings, hence on the norms $\|\xi 1,2\|$. Note the clear dominance of the $\|a\hat{}3\|\nu$ which reflects the fact that $\xi 2=(0,0,1)T$

The above considerations can serve as a starting point for more elaborate future investigations. One might, for example, devise an optimization scheme in which one takes not only the radius *r* as an unknown in the radii polynomials but also considers vv or ||ξ1,2||, respectively, as variables.

5.1.2 Analysis for the Local Lorenz Manifold: Double Eigenvalues

In order to analyze the situation for double eigenvalues as discussed in Sect. 2.3, we choose for the parameters in the Lorenz system (59) the relation ρ=1+(σ+1)2/(4σ), leading to double eigenvalues λ=−(σ+1)/2. Using this data we set up the operator fdoublefdouble specified in (40). Note that the (generalized) eigenvectors ξ1,2ξ1,2 fulfilling (23) can be computed explicitly in this model case. For the general case we refer to methods developed in Alefeld and Spreuer (1986) and Rump (2001). It will be subject of future work to discuss their applicability in the current context. Let us now discuss the influence of the choice of parameters *m* and vv (or ||ξ1,2||||ξ1,2||) we first define the radii polynomials as we did in 5.1.1.

To compute the bounds Yj (j=1,2,3) defined in (54) and Zj (j=1,2,3) defined in (55) we follow the same strategy as in Sect. 5.1.1. First the derivation of the *Y*-bounds is exactly analogous. In deriving zk,i (r), i=1,2i=1,2 fulfilling in the analogue of (61) we notice that the only difference from the formulas in Table 1 is induced by the additional linear term $(k_1+1)a_{\bar{k}}$, with $\bar{k} \stackrel{\text{def}}{=} (k_1+1, k_2-1)$ in fdoublefdouble for |k|≥2| and k2>0. For k2=0 one should read The resulting differences in zk,1 can be read off in Table 3.

Table 3. Expansion coefficients for (Df(x^+rv)rw−A†rw)k for the double eigenvalue case, using the standard Kronecker δ symbol

	$k \in \mathcal{I}_m$	$k \notin \mathcal{I}_m$
$z_{k,1}$	$\delta_{k_1 m_1}(m_1+1)w_{\bar{k}}$	$\begin{pmatrix} \sigma(w^2_{\bar{k}} - w^1_{\bar{k}}) \\ \rho w^1_{\bar{k}} - w^2_{\bar{k}} - ((\hat{a}^1 * w^3)_{\bar{k}} + (\hat{a}^3 * w^1)_{\bar{k}}) \\ (\hat{a}^1 * w^2)_{\bar{k}} + (\hat{a}^2 * w^1)_{\bar{k}} - \beta w^3_{\bar{k}} \end{pmatrix} + (k_1+1)w_{\bar{k}}$

To obtain Zk(r) fulfilling the equivalent of (62) we obtain, in addition to ϵkϵk from (63) and χ2 from (64), a vector χ1∈R3nm such that

$$|z_{k,1}^j| \leq \chi_1^{3(k_2(m_1+1)+k_1)+j} \qquad j = 1, 2, 3.$$

To be precise, we set $\chi_1^{3(k_2(m_1+1)+m_1)+j} = (m_1+1)\nu_1^{-(m_1+1)}\nu_2^{-(k_2-1)}$ for all $1 \leq k_2 \leq m_2$ and $j = 1, 2, 3,$ whereas all other components of χ1χ1 vanish. To obtain the analogue of Z1Z1 in (65) we need to bound the sum

$$\sum_{k \notin \mathcal{I}_m} \frac{k_1+1}{|\lambda_1|(k_1+k_2)} |w_{\bar{k}}^j| \nu_1^{k_1} \nu_2^{k_2} = \frac{\nu_2}{\nu_1} \sum_{k \notin \mathcal{I}_m} \frac{k_1+1}{|\lambda_1|(k_1+k_2)} |w_{\bar{k}}^j| \nu_1^{k_1+1} \nu_2^{k_2-1}.$$

The following lemma can be used to control the factor k1+1|λ1|(k1+k2) uniformly for k∉Imk∉Im. We formulate it in this more general form as it will be reused for more general terms of this type in analyzing resonant eigenvalues in Sect. 5.2.

Lemma 5.1

Let m∈Nd with mı~≥1 for an ı~∈{1,...,d} Let

Qı~= def min{|λi|(mi+1):i≠ı~}.

Then we have, with Im={k∈Nd:k≤m},

$$\max_{k \in \mathbb{N}^d \setminus \mathcal{I}_m} \frac{k_{\bar{\imath}}+1}{|\lambda_1|k_1 + \ldots + |\lambda_d|k_d} \leq \max\left\{ \frac{m_{\bar{\imath}}+2}{|\lambda_{\bar{\imath}}|(m_{\bar{\imath}}+1)}, \frac{m_{\bar{\imath}}+1}{Q_{\bar{\imath}} + |\lambda_{\bar{\imath}}|m_{\bar{\imath}}}, \frac{1}{Q_{\bar{\imath}}} \right\}. \quad (66)$$

Proof 5.1

For any k∈Imk∈Im there is at least one j∈{1,...,d}j∈{1,...,d} such that kj≥mj+1. We distinguish two cases:

$k_{\bar{\imath}} \geq m_{\bar{\imath}} + 1.$ *Then*

- $$\frac{k_{\bar{\imath}}+1}{|\lambda_1|k_1 + \ldots + |\lambda_d|k_d} \leq \frac{k_{\bar{\imath}}+1}{|\lambda_{\bar{\imath}}|k_{\bar{\imath}}} \leq \frac{m_{\bar{\imath}}+2}{|\lambda_{\bar{\imath}}|(m_{\bar{\imath}}+1)}.$$

- $0 \le k_{\bar{i}} \le m_{\bar{i}}$. *We estimate*

$$\frac{k_{\bar{i}}+1}{|\lambda_1|k_1+\ldots+|\lambda_d|k_d} \le \frac{k_{\bar{i}}+1}{Q_{\bar{i}}+|\lambda_{\bar{i}}|k_{\bar{i}}} \le \max\left\{\frac{m_{\bar{i}}+1}{Q_{\bar{i}}+|\lambda_{\bar{i}}|m_{\bar{i}}}, \frac{1}{Q_{\bar{i}}}\right\}.$$

The proof follows from combining these estimates. □⍰

Following the same steps as in Sect. 5.1.1 we obtain Zk,1 as given in Table 4, while Zk,2 remain unaltered from Table 2.

Table 4. Expansion coefficients for $Z_k(r) = Z_{k,1}r + Z_{k,2}r^2$. The absolute value |Am||Am| is to be understood component-wise

	$k \in \mathcal{I}_m$	$k \notin \mathcal{I}_m$
$Z_{k,1}$	$\epsilon_k + (\lvert A_m\rvert \chi_1)_k$	$\frac{1}{\lvert\lambda_1\rvert(k_1+k_2)}\begin{pmatrix}\sigma(\lvert w\rvert^2_k+\lvert w\rvert^1_k)\\ \rho\lvert w\rvert^1_k+\lvert w\rvert^2_k+((\lvert\hat{a}\rvert^1 * \lvert w\rvert^3)_k+(\lvert\hat{a}\rvert^3 * \lvert w\rvert^1)_k)\\ (\lvert\hat{a}\rvert^1 * \lvert w\rvert^2)_k+(\lvert\hat{a}\rvert^2 * \lvert w\rvert^1)_k+\beta\lvert w\rvert^3_k\end{pmatrix} + \frac{k_1+1}{\lvert\lambda_1\rvert(k_1+k_2)}\begin{pmatrix}\lvert w\rvert^1_{\bar{k}}\\ \lvert w\rvert^2_{\bar{k}}\\ \lvert w\rvert^3_{\bar{k}}\end{pmatrix}$

Hence, one finds (in all the cases discussed below) that Z1 is given by

$$Z_1 = \sum_{k\in\mathcal{I}_m} Z_{k,1}\nu^k + \frac{1}{|\lambda_1|\min\{m_1+1, m_2+1\}}\begin{pmatrix} 2|\sigma| \\ |\rho|+1+\|\hat{a}^1\|_\nu+\|\hat{a}^3\|_\nu \\ |\beta|+\|\hat{a}^1\|_\nu+\|\hat{a}^2\|_\nu \end{pmatrix} + \frac{\nu_2}{\nu_1}\frac{m_1+2}{|\lambda_1|(m_1+1)}.$$

Recalling the necessary condition Zj1≤1 from above, we point out that the term ν2ν1m1+2|λ1|(m1+1) is crucial. In particular, we certainly need ν2ν1|λ1|<1ν2ν1|λ1|<1. Therefore, in our analysis of the computational parameters we fix ‖ξ1,2‖ and vary ν1,2.

Remark 5.2

The condition from Lemma 2.7 reads ℓ0(ν^1,ν^2|λ1|)≤ν1. However, the necessary condition ν2ν1|λ1|<1implies that ℓ0(ν1,ν2|λ1|)=ν1, hence in practice one simply takes ν^=νν^=ν.

Let us now consider different choices of parameters β,ρ,σβ,ρ,σ. We fix β=83β=83 and
consider σ=−3,−5,−11,−21 with ρ=43,95,3611,12121ρ=43,95,3611,1212 1. This corresponds to double eigenvalues λ1=1,2,5,10. Thus we compute 2D unstable manifolds (simply replacing *g* by −g everywhere in the analysis). Note that this does not necessitate an adaptation of the formulas for the radii polynomials. Table 5 offers a comparison of the validation success in terms of the decay rate νν.

Table 5. For fixed m=(7,6) and ξ1,2‖=150, we give the corresponding validation radii (if there was one; failure is denoted by --) on the different domain sizes corresponding to the different decay rates νν

	$\lambda_1 = 1$	$\lambda_1 = 2$	$\lambda_1 = 5$	$\lambda_1 = 10$
$\nu = (7, 0.5)$	2.61×10^{-8}	2.58×10^{-12}	1.53×10^{-13}	1.38×10^{-14}
$\nu = (7, 1)$	–	2.06×10^{-11}	4.96×10^{-13}	2.40×10^{-14}
$\nu = (10, 2)$	–	–	2.66×10^{-10}	8.11×10^{-12}
$\nu = (12, 3)$	–	–	–	2.84×10^{-10}

Thus we see that the bigger we choose the magnitude of the eigenvalue, the bigger is the domain of convergence of the parametrization that we are able to validate. Moreover, the smaller the eigenvalue, the bigger the ratio ν1/ν2ν1/ν2 needs to be. This reflects the crucial role of the term ν2ν1m1+2|λ1|(m1+1).

5.2 Coexistence of Hexagonal and Trivial Patterns

As an example of an application to a case with regular resonant eigenvalues, we turn our attention to the existence of solutions of (2) exhibiting certain patterns. The asymptotic analysis in Doelman et al. (2003) reduces the problem of finding transition layers between stationary patterns of (2) to connecting orbit problems for the system of ODEs (1). We proceed to analyze (1) along the same lines as in van den Berg et al. (2015). To this end we set u1=u,u2=u′,u3=vu1=u,u2=u′,u3=v and u4=v′u4=v′ in (1) and rewrite it as

$$u' = g(u) = \begin{pmatrix} u_2 \\ -\frac{\gamma}{4}u_1 - \frac{\sqrt{2}}{4}u_3^2 + \frac{3}{8}u_1^3 + 3u_1u_3^2 \\ u_4 \\ -\gamma u_3 - \frac{\sqrt{2}}{2}u_1u_3 + 9u_3^3 + 3u_1^2u_3 \end{pmatrix} \tag{67}$$

with u=(u1,u2,u3,u4)T. The relation between the parameters γ,μγ,μ and ββ is given by γ=μβ2γ=μβ2. We refer to Doelman et al. (2003) for a complete description of the seminal asymptotic reduction of the PDE (2) to the ODE system (67), obtained via a combination of spatial dynamics, bifurcation theory and geometric singular perturbation theory. For further details about the particular form (67) one may consult (van den Berg et al.2015). We take the same viewpoint as in van den Berg et al. (2015) and investigate by rigorous numerical techniques, for fixed parameter value γγ, connecting orbits between equilibria of (67). While in van den Berg et al. (2015) the configuration for the coexistence of hexagons and rolls (spots and stripes) is considered, we concern ourselves with the coexistence of the trivial state and hexagonal patterns. In terms of (67) this corresponds to a connecting orbit between the two fixed points p1=(2u3∗,0,u3∗,0)T, where u∗3=1+1+60γ√30u3∗=1+1+60γ30, and p2=(0,0,0,0)T. As (67) is Hamiltonian, a necessary condition for the connection between p1p1 and p2 to exist is for the equilibria to lie on the same energy level, which is the case for γ=−2135.

The equilibrium p2 has resonant eigenvalues (λ1=2λ2). For this reason the case of coexistence of the trivial state and the hexagonal spot pattern was not considered in van den Berg et al. (2015). However, our current approach to validating the (un)stable manifold is well suited for this situation.

To compute the connecting orbit we adapt the approach of van den Berg et al. (2015). Introducing a rescaling factor L>0 we aim at solving

$$\begin{cases} u'(t) = Lg(u(t)) & t \in [0,1] \\ u(0) \in W^u_{\text{loc}}(p_1) \\ u(1) \in W^s_{\text{loc}}(p_2) \end{cases} \tag{68}$$

with u(t)=(u1(t),u2(t),u3(t),u4(t))u(t)=(u1(t),u2(t),u3(t),u4(t)). The main aim of the current paper is to construct efficient rigorously validated descriptions of the local (un)stable manifolds. As explained in detail in van den Berg et al. (2015), once we have such parametrizations Pu:R2⊃Bsym1→R4Pu:R2⊃B1sym→R4 of the local unstable manifold ofp1p1 and Ps:R2⊃Bsym1→R4 of the local stable manifold of p2, (68) can be solved by finding a zero of the operator

$$F(\psi, \phi, u)(t) = \begin{pmatrix} u_1(1) - P_1^s(\theta(\psi)) \\ u_3(1) - P_3^s(\theta(\psi)) \\ u_4(1) - P_4^s(\theta(\psi)) \\ u(t) - P^u(\phi) - \int_0^t Lg(u(s))ds \end{pmatrix}, \quad (69)$$

where θ(ψ)= def(ρcos(ψ),ρsin(ψ)) with some fixed ρ<1 (playing the role of the phase condition in this otherwise autonomous problem). The fact that the second component 2(1)−P2s(θ(ψ))=0 can be replaced by the a posteriori check of

$$\text{sign}(u_2(1)) = \text{sign}(P_2^s(\theta(\psi)))$$

is explained in Lemma 2 in van den Berg et al. (2015) and is related to the Hamiltonian nature of the problem. Using the phase condition (i.e., fixing ρρ) and omitting the second component deals on the one hand with the fact that every time shift of a connecting orbit is again a connecting orbit and on the other hand with the fact that the intersection of the two-dimensional unstable and stable manifolds corresponding to the connecting orbit is not transverse in R4R4. In a nutshell, they guarantee the isolation of the zero of *F* that we set out to find.

To rigorously compute zeros of (69) we choose to discretize the time dependence using a *Chebyshev* series (other choices, such as splines, are also possible). For details on how this is done in this example we refer the reader to van den Berg et al. (2015) and for the general method to Lessard and Reinhardt (2014). In this paper we focus on the rigorous computation of the maps Ps,uPs,u and especially PsPs, as there we encounter eigenvalue resonances. The validated computation of PuPu is conducted analogously to the Lorenz equation explained above and we do not give any further details below. Furthermore, in order to validate zeros of (69) we need rigorous information on ‖DPs,u(θ)‖ that we obtain in the same way as explained in Remark 3 in van den Berg et al. (2015). Note that by Remark 4.2 the a posteriori bound δs,uδs,u corresponds to our validation radius *r*.

We now delve into the validated computation of the stable manifold of the origin, which has resonant eigenvalues. Explicitly, the eigenvalues of Dg(p2) with negative real part are given by

$\lambda 1=-−\gamma$ and $\lambda 2=-−\gamma\sqrt{2}\lambda 2=-−\gamma 2$ with corresponding eigenvectors $\xi 1=(0,0,-1-\gamma,1)T$ and $\xi 2=(-2-\gamma,1,0,0)T$. We note that $2\lambda 2=\lambda 1$, so condition (5) holds with ı~=1ı~=1 and k~=(0,2).

Following the approach described in Sect. 2.4 we choose the nonlinear normal form

$$h^s(\theta) = \begin{pmatrix} 2\lambda_2\theta_1 + \tau\theta_2^2 \\ \lambda_2\theta_2 \end{pmatrix} \tag{70}$$

to describe the dynamics in the (stable) parameter space. The coefficients ck=ck(a;γ) in (41) are given by

$$c_k = \begin{pmatrix} a_k^2 \\ -\frac{\gamma}{4}a_k^1 - \frac{\sqrt{2}}{4}(a^3 * a^3)_k + \frac{3}{8}(a^1 * a^1 * a^1)_k + 3(a^1 * a^3 * a^3)_k \\ a_k^4 \\ -\gamma a_k^3 - \frac{\sqrt{2}}{2}(a^1 * a^3)_k + 9(a^3 * a^3 * a^3)_k + 3(a^1 * a^1 * a^3)_k \end{pmatrix}.$$

This completes the ingredients for (41) in the particular case of (67). Note that the condition in Lemma 2.7c) reads ℓ0(θ1,τθ22|λ1|), which we easily check explicitly.

Derivation of the bounds To define the radii polynomials specified in (56b) we first need to compute the bounds *Y* and *Z* defined in (54) and (55). Let f=fregresf=fregres as specified in (41) where ζ=11−γ(0,0,−−γ,1)T∈ker(AT), with *A* defined in (29). Finally, let an approximate solution x^=ιx^F with fm(x^F)≈0be given, and let Am≈(Dfm(x^F))−1. Concerning the bounds *Y* we (again) note that f(x^)k=0f(x^)k=0 for k∉I3m, since *g* is a cubic nonlinearity. Therefore the construction of *Y* fulfilling (54) is analogous to the Lorenz case in Sect. 5.1. Again as in Sect. 5.1, to obtain Zj(r)=Z1jr+Z2jr2+Z3jr3 in (55) we first compute zk(r)=zk,1r+zk,2r2+zk,3r3 fulfilling the analogue of (61). The result is summarized in Table 6 in “Appendix 1.” Note that z−1(r)=0z−1(r)=0, as f−1f−1 is linear in ak~ and mk~≤m for our choice of *m*.

The main deviation from the non-resonant case is given by the off-diagonal linear terms introduced by the resonance. Note that this is analogous to the double eigenvalue case, but with the additional difference given by the presence of the additional unknown ττ. To expand the bounds

$$|(DT(\hat{x}+rv)rw)_k^j| \le Z_{k,1}^j r + Z_{k,2}^j r^2 + Z_{k,3}^j r^3 \qquad \text{for } k \in \mathcal{I}_m = \{0\} \sqcup \{k \in \mathbb{N}^2 : k \preceq m\},$$

we first find bounds ϵkϵk for k∈Imk∈Im such that

$$|((\mathrm{Id} - AA^\dagger)w)_k^j| \le \epsilon_k^j \qquad j = 1,2,3,4. \quad (71)$$

Note that ((Id −AA†)w)k=0∈R4((Id −AA†)w)k=0∈R4 for k∉Imk∉Im by construction. Furthermore we construct, for i=1,2,3i=1,2,3, the vector χi∈R4nm+1 with nm=(m1+1)(m2+1) such that

$$|z_{k,i}^j| \le \chi_i^{1+4(k_2(m_1+1)+k_1)+j} \qquad j = 1,2,3,4. \quad (72)$$

Note that χ1i=0χi1=0 for i=1,2,3i=1,2,3, since f−1f−1 is linear. Then, applying the definition of *A* given in (51) yields the values for Zk,2 and Zk,3 summarized in Table 7 in "Appendix 1." Summing up Zk,I component-wise while splitting into a finite part and infinite tail yields *Z*(*r*). The result is shown in Table 8 in "Appendix 1." An important ingredient in computing Z1 is the control of terms of the form

$$\hat{\tau} \sum_{k \notin \mathcal{I}_m} \frac{k_{\bar{i}}}{|\lambda_1|k_1 + |\lambda_2|k_2} |w_{\underline{k}}| \nu_1^{k_1} \nu_2^{k_2}, \qquad \text{where } \underline{k} \stackrel{\text{def}}{=} (k_1+1, k_2-2). \quad (73)$$

We use Lemma 5.1 to bound the factor |k1+|λ2|k2 uniformly for k∉Im. In the specific problem under consideration we are in luck, since τ^≈0, hence the contribution from the term (73) is small. This stems from the fact that the exact solution is τ=0τ=0 as we derive analytically in "Appendix 2."

However, the strength of our method is that one does *not* need to determine the coefficients of the normal form beforehand, as they are part of the overall set of unknowns for the nonlinear problem.

Numerical implementation The implementation of the validation of an approximate zero of (69) can be found at the webpage Code page

(2015). There, a complete instruction on how to run the codes can be found.

We list the main parameters for the computations, which were chosen after numerical experimentation. As time rescaling parameter we choose $L=43.2034$ in (68). For both the stable and unstable manifold we choose as domain radius in (6) $\nu=(1,1)$.

- Parametrization order: unstable $m=(15,15)m=(15,15)$, stable $m=(15,15)$.
- Scaling of eigenvectors: unstable $\xi 2\|2=0.02$ stable $\|\xi 1\|2=0.0085$, $\|\xi 2\|2=0.0214\|\xi 2\|2=0.0214$

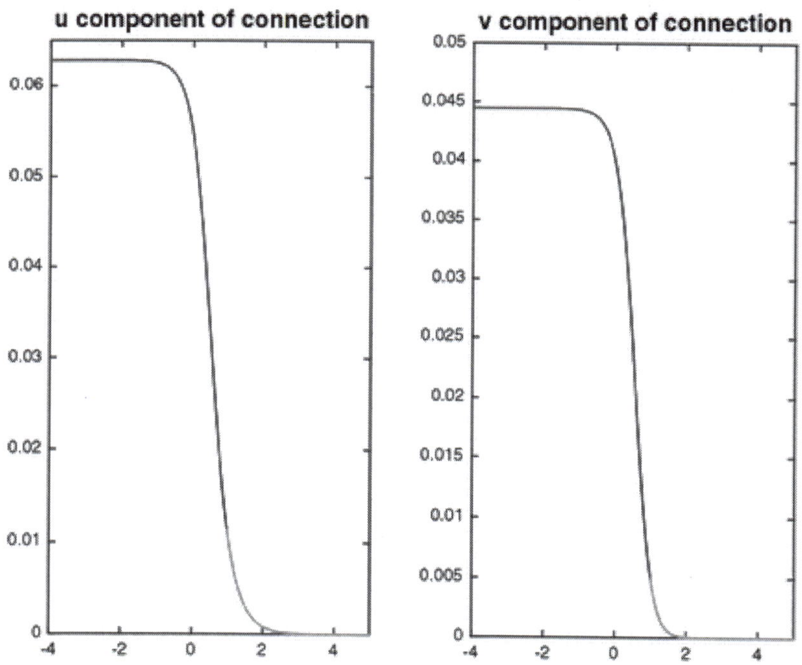

Figure 4. Profiles of the *u* and *v* component of the rigorously verified connecting orbit. The blue part of either orbit is computed using Chebyshev series with the implementation from van den Berg and Sheombarsing (2015) and the red and green parts are computed using the conjugation maps Pu,sPu,s harnessing the formula $u(t)=P(\Psi u,s(t,\theta))$ with $\Psi u,s(t,\theta)$ the flow induced by hu,shu,s for $\theta \in B\hat{}(1,1)sym$ (see Lemma 2.6). One time unit corresponds to $L=43.2034$.

The *Y*-bounds are explained above and the *Z*-bounds which we use to define the radii polynomials from (56) are summarized in Table 8. Using them to check the conditions of Lemma 4.6 we obtain the following validation radii:

unstable manifold:r=1.0823×10−12,stable manifold:r=5.2217×10−12.unstable manifold:r=1.0823×10−12,stable manifold:r=5.2217×10−12.

For the validation of the connection we use the approach of van den Berg et al. (2015) in conjunction with the implementation of van den Berg and Sheombarsing (2015). The validated solution profiles are shown in Fig. 4.

Appendix 1: Formulas for the Expansion Coefficients zk and Zk for (67)

Let f=fregresf=fregres defined in (41) be initiated according to the resonance present in the stable manifold of the origin in R4R4 as a fixed point of (67). That is k~=(0,2)k~=(0,2) and ı~=1 in (41). Let x^=(τ^,a^)=(τ^,a^1,a^2,a^3,a^4). To make the expansion

$$(Df(\hat{x}+rv)rw - A^{\dagger}rw)_k = z_{k,1}r + z_{k,2}r^2 + z_{k,3}r^3$$

explicit, we first define the operations

$$H_{ij}(a) = a^i * a^j \qquad \text{and} \qquad H_{ijl}(a) = a^i * a^j * a^l.$$

Denote v=(τv,va) and v=(τw,wa). By a direct calculation one derives expansions of the form

$$H_{ij}(\hat{a}+rv)rw = \eta_{ij,1}r + \eta_{ij,2}r^2$$
$$H_{ijl}(\hat{a}+rv)rw = \eta_{ijl,1}r + \eta_{ijl,2}r^2 + \eta_{ijl,3}r^3,$$

where, for example, ηij,2=vai∗waj+vaj∗wai.

Recall that the special case k=−1k=−1 yields (Df(x^+rv)rw−A†rw)−1=0. The result for the remaining*k* is given in Table 6, where we use the notation k−=(k1+1,k2−2), and whenever k2<2 one should read 0 for terms involving k−k_. Concerning the bounds χi defined in (72), we proceed as in Sect. 5.1. In order to define χ1 we use

$$|w_{\underline{k}}^{j}| \leq \frac{1}{\nu_1^{k_1+1}\nu_2^{k_2-2}}$$

to control the term (73) for k1=m1. Note that all other components of χ1 vanish. To define χ2,3χ2,3 we note, in addition, that

$$|(\eta_{ij,2})_{\underline{k}}^{j}| \leq \frac{2}{\nu^k}, \qquad |(\eta_{ijl,2})_{\underline{k}}^{j}| \leq \frac{2}{\nu^k}\left(\|\hat{a}^i\|_\nu + \|\hat{a}^j\|_\nu + \|\hat{a}^l\|_\nu\right), \qquad |(\eta_{ijl,3})_{\underline{k}}^{j}| \leq \frac{3}{\nu^k}.$$

In Table 7 we summarize the coefficients Zk,i appearing in the bounds (DT(x^+rv)rw)kj|≤Zk,1jr+Zk,2jr2+Zk,3jr3. We are now ready to define Z−1(r)=Z−1,1r

Table 6. Expansion coefficients for $(Df(\hat{x}+rv)rw - A^\dagger rw)_k$

	$k \in \mathcal{I}_{0,m}$	$k \notin \mathcal{I}_{1,m}$
$z_{k,1}$	$\delta_{k_1,m_1}\hat{\tau}(m_1+1)w_{\underline{k}}$	$\begin{pmatrix} w_{\underline{k}}^2 \\ -\gamma w_{\underline{k}}^1 - \frac{\sqrt{2}}{4}(\eta_{33,1})_k + \frac{3}{8}(\eta_{111,1})_k + 3(\eta_{133,1})_k \\ w_{\underline{k}}^4 \\ -\gamma w_{\underline{k}}^3 - \frac{\sqrt{2}}{2}(\eta_{13,1})_k + 9(\eta_{333,1})_k + 3(\eta_{113,1})_k \end{pmatrix} + \hat{\tau}(k_1+1)w_{\underline{k}}$
$z_{k,2}$	$\begin{pmatrix} 0 \\ -\frac{\sqrt{2}}{4}(\eta_{33,2})_k + \frac{3}{8}(\eta_{111,2})_k + 3(\eta_{133,2})_k \\ 0 \\ -\frac{\sqrt{2}}{2}(\eta_{13,2})_k + 9(\eta_{333,2})_k + 3(\eta_{113,2})_k \end{pmatrix} + (k_1+1)(v_{-1}w_{\underline{k}} + w_{-1}v_{\underline{k}})$	$\begin{pmatrix} 0 \\ -\frac{\sqrt{2}}{4}(\eta_{33,2})_k + \frac{3}{8}(\eta_{111,2})_k + 3(\eta_{133,2})_k \\ 0 \\ -\frac{\sqrt{2}}{2}(\eta_{13,2})_k + 9(\eta_{333,2})_k + 3(\eta_{113,2})_k \end{pmatrix} + (k_1+1)(v_{-1}w_{\underline{k}} + w_{-1}v_{\underline{k}})$
$z_{k,3}$	$\begin{pmatrix} 0 \\ \frac{3}{8}(\eta_{111,3})_k + 3(\eta_{133,3})_k \\ 0 \\ 9(\eta_{333,3})_k + 3(\eta_{113,3})_k \end{pmatrix}$	$\begin{pmatrix} 0 \\ \frac{3}{8}(\eta_{111,3})_k + 3(\eta_{133,3})_k \\ 0 \\ 9(\eta_{333,3})_k + 3(\eta_{113,3})_k \end{pmatrix}$

The explicit expression for Zj(r)=Z1jr+Z2jr2+Z3jr3, j=1,2,3,4j=1,2,3,4 is collected in Table 8.

Table 7. Bounds for $|(DT(\hat{x}+rv)rw)_{\underline{k}}^{j}|$

	$k \in \mathcal{I}_{1,m}$	$k \notin \mathcal{I}_{1,m}$
$Z_{k,1}$	$\epsilon_k + (\lvert A_m\rvert\chi_1)_k$	$\frac{1}{k_1\lvert\lambda_1\rvert+k_2\lvert\lambda_2\rvert}\begin{pmatrix} \lvert w_{\underline{k}}^2\rvert \\ \lvert\gamma\rvert\lvert w_{\underline{k}}^1\rvert + \frac{\sqrt{2}}{4}\lvert(\eta_{33,1})_k\rvert + \frac{3}{8}\lvert(\eta_{111,1})_k\rvert + 3\lvert(\eta_{133,1})_k\rvert \\ \lvert w_{\underline{k}}^4\rvert \\ \gamma\lvert w_{\underline{k}}^3\rvert + \frac{\sqrt{2}}{2}\lvert(\eta_{13,1})_k\rvert + 9\lvert(\eta_{333,1})_k\rvert + 3\lvert(\eta_{113,1})_k\rvert \end{pmatrix} + \frac{\lvert\hat{\tau}\rvert(k_1+1)}{k_1\lvert\lambda_1\rvert+k_2\lvert\lambda_2\rvert}\lvert w_{\underline{k}}\rvert$
$Z_{k,2}$	$(\lvert A_m\rvert\chi_2)_k$	$\frac{1}{k_1\lvert\lambda_1\rvert+k_2\lvert\lambda_2\rvert}\begin{pmatrix} 0 \\ \frac{\sqrt{2}}{4}\lvert(\eta_{33,2})_k\rvert + \frac{3}{8}\lvert(\eta_{111,2})_k\rvert + 3\lvert(\eta_{133,2})_k\rvert \\ 0 \\ \frac{\sqrt{2}}{2}\lvert(\eta_{13,2})_k\rvert + 9\lvert(\eta_{333,2})_k\rvert + 3\lvert(\eta_{113,2})_k\rvert \end{pmatrix} + \frac{\lvert(k_1+1)}{k_1\lvert\lambda_1\rvert+k_2\lvert\lambda_2\rvert}(\lvert w_{\underline{k}}\rvert + \lvert v_{\underline{k}}\rvert)$
$Z_{k,3}$	$(\lvert A_m\rvert\chi_3)_k$	$\frac{1}{k_1\lvert\lambda_1\rvert+k_2\lvert\lambda_2\rvert}\begin{pmatrix} 0 \\ \frac{3}{8}\lvert(\eta_{111,3})_k\rvert + 3\lvert(\eta_{133,3})_k\rvert \\ 0 \\ 9\lvert(\eta_{333,3})_k\rvert + 3\lvert(\eta_{113,3})_k\rvert \end{pmatrix}$

Table 8. *Z*-bounds

Z_1	$\sum_{k\in\mathcal{I}_{0,m}}\begin{pmatrix}\lvert Z^1_{k,1}\rvert\\ \lvert Z^2_{k,1}\rvert\\ \lvert Z^3_{k,1}\rvert\\ \lvert Z^4_{k,1}\rvert\end{pmatrix}\nu^k+\frac{1}{\min\{\lvert\lambda_1\rvert(m_1+1),\lvert\lambda_2\rvert(m_2+1)\}}\begin{pmatrix}1\\ \lvert\gamma\rvert+\sqrt{2}\lVert\hat a^3\rVert_\nu+\frac{9}{8}\lVert\hat a^1\rVert_\nu^2+3\left(2\lVert\hat a^1\rVert_\nu\lVert\hat a^3\rVert_\nu+\lVert\hat a^3\rVert_\nu^2\right)\\ 1\\ \lvert\gamma\rvert+\frac{\sqrt{2}}{2}\left(\lVert\hat a^1\rVert_\nu+\lVert\hat a^3\rVert_\nu\right)+27\lVert\hat a^3\rVert_\nu^2+3\left(2\lVert\hat a^1\rVert_\nu\lVert\hat a^3\rVert_\nu+\lVert\hat a^1\rVert_\nu^2\right)\end{pmatrix}+\frac{\lvert\hat\tau\rvert\nu_2^2}{\nu_1}\frac{m_1+1}{\lvert\lambda_1\rvert m_1}\begin{pmatrix}1\\1\\1\\1\end{pmatrix}$
Z_2	$\sum_{k\in\mathcal{I}_{0,m}}\begin{pmatrix}\lvert Z^1_{k,2}\rvert\\ \lvert Z^2_{k,2}\rvert\\ \lvert Z^3_{k,2}\rvert\\ \lvert Z^4_{k,2}\rvert\end{pmatrix}\nu^k+\frac{1}{\min\{\lvert\lambda_1\rvert(m_1+1),\lvert\lambda_2\rvert(m_2+1)\}}\begin{pmatrix}0\\ \frac{\sqrt{2}}{2}+\frac{54}{8}\lVert\hat a^1\rVert_\nu+3\left(2\lVert\hat a^1\rVert_\nu+4\lVert\hat a^3\rVert_\nu\right)\\ 0\\ \frac{\sqrt{2}}{2}+54\lVert\hat a^3\rVert_\nu+3\left(4\lVert\hat a^1\rVert_\nu+2\lVert\hat a^1\rVert_\nu^2\right)\end{pmatrix}+\frac{\nu_2^2}{\nu_1}\frac{m_1+1}{\lvert\lambda_1\rvert m_1}\begin{pmatrix}2\\2\\2\\2\end{pmatrix}$
Z_3	$\sum_{k\in\mathcal{I}_{0,m}}\begin{pmatrix}\lvert Z^1_{k,3}\rvert\\ \lvert Z^2_{k,3}\rvert\\ \lvert Z^3_{k,3}\rvert\\ \lvert Z^4_{k,3}\rvert\end{pmatrix}\nu^k+\frac{1}{\min\{\lvert\lambda_1\rvert(m_1+1),\lvert\lambda_2\rvert(m_2+1)\}}\begin{pmatrix}0\\ \frac{81}{8}\\ 0\\ 36\end{pmatrix}$

Appendix 2: Analytical Proof that τ=0.

In the following we show analytically that the additional parameter ττ in (70) vanishes for the exact solution.

Imposing the linear constraints (9) we obtain

$$a_{00}=(0,0,0,0)^T\quad a_{10}=(0,0,-\frac{1}{\sqrt{-\gamma}},1)^T\quad a_{01}=(-\frac{2}{\sqrt{-\gamma}},1,0,0)^T. \tag{74}$$

In order to show that τ=0 we use the formula (31). Recall that we computed ζ=11−γ(0,0,−−γ,1)T. Considering (29) with ı~=1 and k~=(0,2) with the equilibrium p=(0,0,0,0) for *g* specified in (67) we compute a concrete formula for b02:

$$b_{02}=\begin{pmatrix}0\\ -\frac{\sqrt{2}}{4}(a^3_{01})^2+\frac{9}{8}a^1_{00}(a^1_{01})^2+3\left(a^3_{00}(a^1_{01})^2+2a^3_{01}a^1_{00}a^1_{01}\right)\\ 0\\ -\frac{\sqrt{2}}{2}a^1_{01}a^3_{01}+27a^3_{00}a^3_{01}a^3_{01}+3\left(a^1_{00}(a^3_{01})^2+2a^1_{01}a^3_{00}a^3_{01}\right)\end{pmatrix}.$$

Using (74) we see that b02=(0,0,0,0), which implies τ=0. As a result one could deal with this particular resonance in an ad-hoc fashion by prescribing in addition to (74) a02 as any solution of (Dg(0)−λ1I4)a02=0 and choosing

$$h^s(\theta)=\begin{pmatrix}\lambda_1\theta_1\\ \lambda_2\theta_2\end{pmatrix}$$

just as in the non-resonant case. The one we singled out by imposing (30) is a02=(0,0,0,0).

REFERENCES

1. Alefeld, G., Spreuer, H.: Iterative improvement of componentwise error bounds for invariant subspaces belonging to double or nearly double eigenvalues. Computing **36**(4), 321–334 (1986)

2. Arioli, G., Koch, H.: Computer-assisted methods for the study of stationary solutions in dissipative systems, applied to the Kuramoto–Sivashinski equation. Arch. Ration. Mech. Anal. **197**(3), 1033–1051 (2010)

3. Arioli, G., Koch, H.: Existence and stability of traveling pulse solutions of the FitzHugh–Nagumo equation. Nonlinear Anal. **113**, 51–70 (2015)

4. Beyn, W.-J., Kless, W.: Numerical Taylor expansions of invariant manifolds in large dynamical systems. Numer. Math. **80**, 1–38 (1998)

5. Breden, M., Lessard, J.P., Mireles James, J.D.: Computation of maximal local (un)stable manifold patches by the parameterization method. (2015, submitted)

6. Breden, M., Lessard, J.P., Vanicat, M.: Global bifurcation diagram of steady states of systems of pdes via rigorous numerics. Acta Appl. Math. **128**(1), 113–152 (2013)

7. Cabré, X., Fontich, E., de la Llave, R.: The parameterization method for invariant manifolds I: manifolds associated to non-resonant subspaces. Indiana Univ. Math. J. **52**, 283–328 (2003)

8. Cabré, X., Fontich, E., de la Llave, R.: The parametrization method for invariant manifolds II: regularity with respect to parameters. Indiana Univ. Math. J. **52**, 283–328 (2003)

9. Cabré, X., Fontich, E., de la Llave, R.: The parametrization method for invariant manifolds III: overview and applications. J. Differ. Equ. **218**(2), 444–515 (2005)

10. Canadell, M., Haro, A.: Parameterization method for computing quasi-periodic reducible normally hyperbolic invariant tori. In: Formaggia, L., Pedregal, P. (eds.) Advances in Differential

Equations and Applications. SEMA SIMAI Springer Series, vol. 4, pp. 85–94 (2014)

11. CAPD: Computer assisted proofs in dynamics, a package for rigorous numerics. 2015.

12. Capinksi, M., Simo, C.: Computer assisted proof for normally hyperbolic invariant manifolds. Nonlinearity **25**(7), 1997–2026 (2012)

13. Castelli, R., Lessard, J.-P., Mireles-James, J.D.: Parametrization of invariant manifolds for periodic orbits (I): efficient numerics via the Floquet normal form. SIAM J. Appl. Dyn. Syst. **14**, 132–167 (2015)

14. Chow, S.-N., Li, C., Wang, D.: Normal Forms and Bifurcation of Planar Vector Fields. Cambridge University Press, Cambridge (1994)

15. Code page. http://www.few.vu.nl/~crt290/manifoldcode.php. Accessed 18 May 2015

16. Day, S., Lessard, J.-P., Mischaikow, K.: Validated continuation for equilibria of PDEs. SIAM J. Numer. Anal.**45**(4), 1398–1424 (2007)

17. Doelman, A., Sandstede, B., Scheel, A., Schneider, G.: Propagation of hexagonal roles near onset. Eur. J. Appl. Math. **14**(1), 85–110 (2003)

18. Eckmann, J.-P., Koch, H., Wittwer, P.: A computer-assisted proof of universality for area-preserving maps. Mem. Amer. Math. Soc. **47**(289), vi+122 (1984)

19. Falcolini, C., de la Llave, R.: Numerical calculation of domains of analyticity for perturbation theories in the presence of small divisors. J. Stat. Phys. **67**(3–4), 645–666 (1992)

20. Govaerts, W.F.: Numerical Methods for Bifurcations of Dynamical Equilibria. SIAM, Philadelphia (2000)

21. Guillamon, A., Huguet, G.: A computational and geometric approach to phase resetting curves and surfaces. SIAM J. Appl. Dyn. Syst. **8**(3), 1005–1042 (2009)

22. Haro, A.: Automatic differentiation methods in computational dynamical systems: invariant manifolds and normal forms of vector fields at fixed points. Notes for the IMA New Directions short course "Invariant objects in dynamical systems and their applications" (2011)

23. Haro, A., Canadell, M., Figueras, J.-L.L., Luque, A., Mondelo, J.-M.: The parameterization method for invariant manifolds: from theory to effective computations, volume, 2014. Preprint.

24. Haro, A., da la Llave, R.: A parametrization method for the computation of invariant tori and their whiskers in quasi-periodic maps: Numerical algorithms. Discrete Contin. Dyn. Syst. Ser. B **6**(6), 1261–1300 (2006)

25. Huguet, G., de la Llave, R.: Computation of limit cycles and their isochrons: fast algorithms and their convergence. SIAM J. Appl. Dyn. Syst. **12**(4), 1763–1802 (2013)

26. Hungria, A., Lessard, J.P., Mireles-James, J.: Rigorous numerics for analytical solutions of differential equations: the radii polynomial approach. Math. Comput. **85**(299), 1427–1459 (2016)

27. Johnson, T., Tucker, W.: A note on the convergence of parametrised non-resonant invariant manifolds. Qual. Theory Dyn. Syst. **10**(1), 107–121 (2011)

28. Krauskopf, B., Osinga, H.M., Doedel, E.J., Henderson, M.E., Guckenheimer, J.M., Vladimirsky, A., Dellnitz, M., Junge, O.: A survey of methods for computing (un)stable manifolds of vector fields. Int. J. Bifurcation Chaos **15**(3), 763–791 (2005)

29. Kuznetsov, YuA: Elements of Applied Bifurcation Theory. Springer, New York (2004)

30. Lanford III, O.E.: A computer-assisted proof of the Feigenbaum conjectures. Bull. Am. Math. Soc. (N.S.) **6**(3), 427–434 (1982)

31. Lessard, J.-P., Reinhardt, C.: Rigorous numerics for nonlinear odes using Chebyshev series. SIAM J. Numer. Anal. **52**(1), 1–22 (2014)

32. Lorenz, E.N.: Deterministic nonperiodic flows. J. Atmos. Sci. **20**, 130–141 (1963)

33. Mireles-James, J.: Quadratic volume-preserving maps: (un)stable manifolds, hyperbolic dynamics and vortex-bubble bifurcations. J. Nonlinear Sci. **23**(4), 585–615 (2013)

34. Mireles-James, J.: Polynomial approximation of a one parameter family of (un)stable manifolds with rigorous error bounds. Indag. Math. **26**, 225–265 (2015)

35. Mireles James, J.D., Lomelí, H.: Computation of heteroclinic arcs with application to the volume preserving Hénon family. SIAM J. Appl. Dyn. Syst. **9**(3), 919–953 (2010)

36. Mireles-James, J., Mischaikow, K.: Rigorous a-posteriori computation of (un)stable manifolds and connecting orbits for analytic maps. SIAM J. Appl. Dyn. Syst. **12**(2), 957–1006 (2013)

37. Rump, S.M.: Computational error bounds for multiple or nearly multiple eigenvalues. Linear Algebra Appl.**324**(1–3), 209–226 (2001)

38. van den Berg, J.B., Sheombarsing, R.: Domain decomposition in rigorous numerics for ODEs. (2016, submitted)

39. van den Berg, J.B., Lessard, J.P., Mischaikow, K.: Global smooth solution curves using rigorous branch following. Math. Comput. **79**(271), 1565–1584 (2010)

40. van den Berg, J.B., Lessard, J.-P., Mireles-James, J., Mischaikow, K.: Rigorous numerics for symmetric connecting orbits: even homoclinics of the Gray–Scott equation. SIAM J. Math. Anal. **43**(4), 1557–1594 (2011)

41. van den Berg, J.B., Mireles-James, J., Deschênes, A., Lessard, J.-P.: Stationary coexistence of hexagons and rolls via rigorous computations. SIAM J. Appl. Dyn. Syst. **14**(2), 942–979 (2015)

42. Wittig, A.: Rigorous high-precision enclosures of fixed points and their invariant manifolds. Ph.D. thesis, Michigan State University (2011)

43. Wittig, A., Berz, M., Grote, J., Makino, K., Newhouse, S.: Rigorous and accurate enclosure of invariant manifolds on surfaces. Regul. Chaotic Dyn. **15**(2–3), 107–126 (2010)

44. Yamamoto, N.: A numerical verification method for solutions of boundary value problems with local uniqueness by Banach's fixed-point theorem. SIAM J. Numer. Anal. **35**(5), 2004–2013 (1998). (electronic)

5

CHAPTER

Traveling Surface Waves of Moderate Amplitude in Shallow Water

Armengol Gasull , Anna Geyer

Departament de Matemàtiques, Facultat de Ciències, Universitat Autònoma de Barcelona, 08193 Bellaterra, Barcelona, Spain

ABSTRACT

We study traveling wave solutions of an equation for surface waves of moderate amplitude arising as a shallow water approximation of the Euler equations for inviscid, incompressible and homogeneous fluids. We obtain solitary waves of elevation and depression, including a family of solitary waves with compact support, where the amplitude may increase or decrease with respect to the wave speed. Our approach is based on techniques from dynamical systems and relies on a reformulation of the evolution equation as an autonomous Hamiltonian system which facilitates an explicit expression for bounded orbits in the phase plane to establish existence of the corresponding periodic and solitary traveling wave solutions.

Keywords: Solitary waves; Homoclinic orbit; Compact support; Shallow water

1. INTRODUCTION AND MAIN RESULT

A number of competing nonlinear model equations for water waves have been proposed to this day to account for fascinating phenomena, such as wave breaking or solitary waves, which are not captured by linear theory. The well-known Camassa–Holm equation [1] is one of the most prominent examples, due to its rich structural properties. It is an integrable infinite-dimensional Hamiltonian system [2], [3] and [4] whose solitary waves are solitons [5] and [6]. Some of its classical solutions develop singularities in finite time in the form of wave breaking [7], and recover in the sense of global weak solutions after blow up [8] and [9]. For a discussion on integrability in the periodic case we refer the reader to [10] and [11], and a classification of weak traveling wave solutions of the Camassa–Holm equation may be found in [12]. The manifold of its enticing features led Johnson to demonstrate the relevance of the Camassa–Holm equation as a model for the propagation of shallow water waves of moderate amplitude. He proved that the horizontal component of the fluid velocity field at a certain depth within the fluid is indeed described by a Camassa–Holm equation [13] and [14]. Constantin and Lannes [15] followed up on the matter in search of a suitable corresponding equation for the free surface and derived an evolution equation for surface waves of moderate amplitude in the shallow water regime,

$$u_t + u_x + 6uu_x - 6u^2u_x + 12u^3u_x + u_{xxx} - u_{xxt} + 14uu_{xxx} + 28u_xu_{xx} = 0. \tag{1}$$

The authors show that Eq. (1) approximates the governing equations to the same order as the Camassa–Holm equation, and also prove that the Cauchy problem on the line associated to (1), is locally well-posed [15]. Employing a semigroup approach due to Kato [16], Duruk [17] shows that this result also holds true for a larger class of initial data, as well as for the corresponding spatially periodic Cauchy problem [18]. Consequently, solutions of (1) depend continuously on their initial data in H^s for $s>3/2$, and it can be shown that this dependence is not uniformly continuous [19]. In the context of Besov spaces, well-posedness is discussed [20] using Littlewood–Paley decomposition, along with a study about analytic solutions and persistence properties of strong solutions. One of the important aspects of Eq. (1) lies in its relevance for capturing the non-linear phenomenon of wave breaking [15] and [17], a feature it shares with the Camassa-Holm equation. While the latter equation is known to possess global solutions [9] and [21], it is not apparent how to obtain global control of the solutions of Eq. (1), owing to its involved structure and due to the higher order nonlinearities. However, passing to a moving frame one can study so-called traveling wave solutions, whose wave profiles move at constant speed in one direction without altering their shape. Introducing the traveling wave Ansatz

$$\xi = x - ct, \qquad u(\xi) = u(t, x), \tag{2}$$

Eq. (1) reads upon integration

$$(1-c)u + 3u^2 - 2u^3 + 3u^4 + (1 + c + 14u)\ddot{u} + 7\dot{u} = E, \tag{3}$$

for some real constant E, where the dot denotes differentiation with respect to ξ. Existence of smooth solitary wave solutions of (3) which decay to zero at infinity has been established [22] for wave speeds $c>1$, and their orbital stability has been deduced [23] employing an approach due to Grillakis, Shatah and Strauss [24] taking advantage of the Hamiltonian structure of (1). In the present paper we set out to improve the existence result [22] by loosening the assumption that solitary waves tend to zero at infinity. Allowing for a decay to an arbitrary constant, we establish existence of a variety of novel traveling wave solutions of (1).

Theorem 1.1.

For every speed $c \in \mathbb{R} \setminus \{c^*\}$ *there exist peaked periodic, as well as smooth solitary and periodic traveling wave solutions of* (1). *Periodic waves may be obtained also for* $c=c^*$, *where* $c^* \approx 0.35328$ *is the unique real solution of* $3c^3+30c^2+1031c-368=0$.

Moreover, the solitary waves can be characterized in terms of two parameters–the wave speed c *and the level of the undisturbed water surface* s*–allowing us to determine the exact regions in this parameter space which give rise to the following types of waves, cf. Fig. 1:*

- *Solitary waves in* C^1 *with* compact support *on* $\mathbb{R}$ *(along the straight line* $c+1+14s=0$*).*
- *Smooth solitary waves of* elevation *whose amplitude is strictly* increasing *(in region* I*) or* decreasing *(in regions* II *and* III*) with respect to* c*, once* E *is fixed.*
- *Smooth solitary waves of* depression *whose amplitude is strictly* increasing *(in region* V*) or* decreasing *(in region* IV*) with respect to* c*, once* E *is fixed.*

All solitary waves are symmetric with respect to their unique crest/trough, they are monotonic and decay exponentially to the undisturbed water surface s *at infinity.*

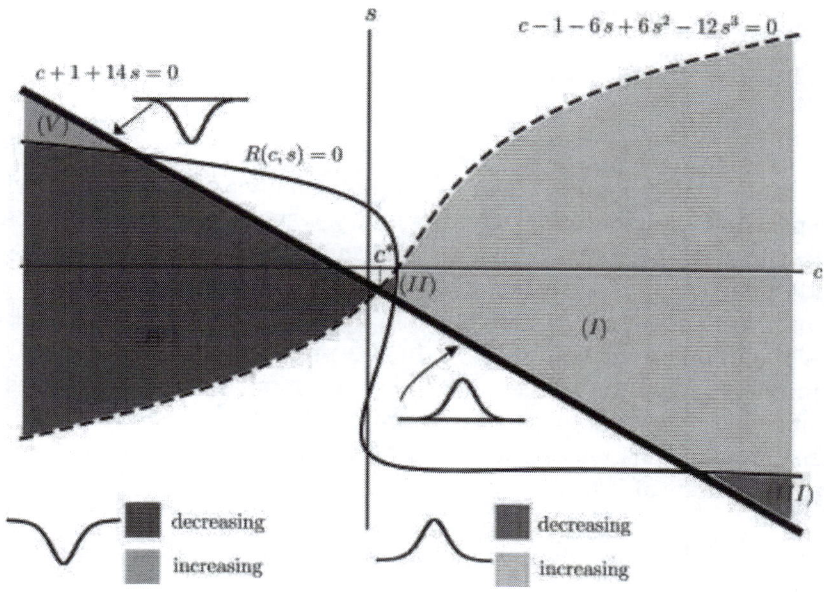

Figure 1. We obtain a variety of traveling waves for (1) with different behavior, cf. Theorem 1.1: solitary waves of *elevation* and *depression* whose amplitude *increases* or *decreases* with c. Along the straight line $c+1+14s=0$, we obtain solitary waves with *compact support* . The algebraic curve $R(c,s)=0$ arising from a polynomial of degree nine is given in Section 3.1.

The proof of these results hinges on the observation that for traveling waves, Eq. (1) may be written as an autonomous Hamiltonian system involving two parameters. This insight allows us to explicitly determine bounded orbits in the phase plane which correspond to solitary and periodic traveling waves of elevation as well as depression (Sections 2.1 and 2.2). Moreover, we characterize all solitary traveling waves in terms of two parameters—the wave speed c and the water level s of the undisturbed surface at infinity. This enables us to prove the existence a family of solitary waves with compact support (Section 2.3). Furthermore, we obtain a family of peaked periodic waves (Section 2.4). Our work also extends the analysis of qualitative properties regarding the shape of solitary waves given in [22]: we prove that the profile is strictly monotonic between crest and trough, and derive explicit algebraic curves in the parameter space (c,s) to determine the regions where their amplitude is increasing and decreasing with respect to the wave speed c once a value E is fixed (Section 3). These are quite remarkable properties of solitary waves which, to our knowledge, contribute novel aspects to the study of traveling waves in evolution equations for water waves. Our approach is in fact applicable to a large class of nonlinear dispersive equations, which we exemplify by a discussion of traveling waves of the aforementioned Camassa–Holm equation, including peaked continuous solitary waves (Section 4). Some of the more involved computations regarding the algebraic curves are carried out in the Appendix.

2. EXISTENCE OF TRAVELING WAVES

The proof of Theorem 1.1 relies on the fact that Eq. (1) has very nice structural features.

2.1. Hamiltonian formulation

Consider a general partial differential equation which, upon introducing the traveling wave Ansatz (2) can be transformed into an autonomous ordinary differential equation of the form

$$\ddot{u}(u-\bar{u})+\frac{1}{2}(\dot{u})^2+F'(u)=0, \tag{4}$$

where $\bar{u}$ is a constant, $F(u)$ is a smooth function and the dot denotes differentiation with respect to ξ. The corresponding planar system is given by

$$\begin{cases} \dot{u} = v \\ \dot{v} = \dfrac{-F'(u) - \frac{1}{2}v^2}{u - \bar{u}}, \end{cases} \tag{5}$$

and we observe that a reparametrization of the independent variable according to $\frac{d\xi}{d\tau} = u - \bar{u}$ transforms (5) into

$$\begin{cases} u' = (u - \bar{u})v \\ v' = -F'(u) - \dfrac{1}{2}v^2, \end{cases} \tag{6}$$

where the prime denotes differentiation with respect to τ. The latter system is clearly topologically equivalent to the former (cf. [25] and [26]) on each connected component of $\mathbb{R}\backslash\{u = \bar{u}\}$, preserving orientation in the open half-plane $\{u > \bar{u}\}$ and reversing orientation in the other half. The advantage of (6) is that it possesses a Hamiltonian

$$H(u, v) = F(u) + \frac{1}{2}v^2(u - \bar{u}) = h \tag{7}$$

satisfying $u' = H_v$ and $v' = -H_u$, which is constant along the solution curves of (6). Explicit knowledge of the critical points and (closed) orbits

$$v = \pm\sqrt{2\frac{h - F(u)}{u - \bar{u}}} \tag{8}$$

in the phase plane of (6) therefore completely characterizes the traveling wave solutions of the partial differential equation. Applying these ideas to (1), the associated equation for travelling waves (3) is

$$(1-c)u + 3u^2 - 2u^3 + 3u^4 + (1 + c + 14u)\ddot{u} + 7\dot{u} = E = -14K,$$

for some constant $K \in R$, which may be written in the form (4) with

$$F(u) = K\,u + \frac{1-c}{28}u^2 + \frac{1}{14}u^3 - \frac{1}{28}u^4 + \frac{3}{70}u^5, \tag{9}$$

and

$$\bar{u} = -\frac{1+c}{14}.$$

In view of the above considerations, we obtain the following existence result for bounded traveling wave solutions of (1):

Proposition 2.1.

Solitary wave solutions of (4) with F defined in (9) are obtained from the homoclinic connection based at the saddle point of (6), whereas periodic waves correspond to periodic orbits around the center. These solutions are symmetric with respect to the crest/trough and have one maximum per period. The solitary waves tend exponentially to a constant on either side of the crest/trough.

Proof.

In order to obtain bounded orbits of system (6), we study the critical points of the Hamiltonian system which are closely related to the local extrema of F, since $(u', v') = (0, 0)$ exactly when $v = 0$ and $F'(u) = 0$. Notice that $u = \bar{u}$ is an invariant line for (6). The fact that any non-degenerate critical point of an analytic Hamiltonian system is either a topological saddle or a center (cf. [27, p. 154]) simplifies our analysis considerably. Computing the Jacobian J of (6) and evaluating it at the critical points shows that $\det J = F''(u)(u - \bar{u})$. Recall that a non-degenerate critical point is a center whenever $\det J > 0$ and a topological saddle when $\det J < 0$, cf. [27]. Hence, all further analysis regarding the number, location and type of critical points in the phase plane is based on the specific structure of the polynomial F, depending on the parameters c and K. It is straightforward to check that F has at most two local extrema and we conclude that system (6) has at most two critical points.

Next we show how to obtain the expressions for bounded orbits corresponding to bounded traveling wave solutions from relation (8), cf. Fig. 2, and infer some basic properties of the waves. Homoclinic orbits are obtained by letting $h = h_s$, where $h_s = F(s)$ and s solves

$$F'(y) = 0 \quad \text{and} \quad \begin{cases} F''(y) < 0 & \text{when } y > \bar{u}, \\ F''(y) > 0 & \text{when } y < \bar{u}. \end{cases} \tag{10}$$

We will see below that this definition of s giving rise to the saddle point $(s, 0)$ of system (6), aptly captures the physical interpretation of s being the level of the undisturbed water surface at infinity in certain regions of the parameter space.

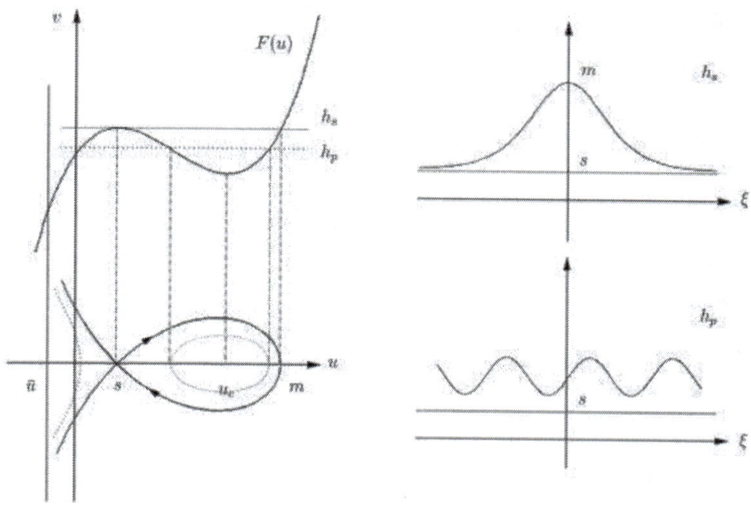

Figure 2. A sketch of how bounded orbits in the (u,v) phase plane are obtained using relation (8), where choosing $h = h_p$ and $h = h_s$ give rise to periodic and solitary traveling waves, respectively.

Regarding v in (8) as a function of u and choosing $h = h_s$ yields an explicit expression of the homoclinic connection in the phase plane. The orbit leaves the critical saddle point $(s, 0)$ and crosses the horizontal axis once at $(m, 0)$, before returning to the saddle point symmetrically with respect to the horizontal axis. The value m is obtained at the unique intersection of the horizontal line h_s with the polynomial F, where $F(m) = F(s)$. The corresponding solitary wave solution therefore has a unique maximum m and is symmetric with respect to the crest point. Furthermore, the solution decays exponentially to the constant s on either side of the crest. Indeed, by the Hartman–Grobman Theorem (cf. [28] and [29]) the vectorfield $(\dot{u}, \dot{v})$ is locally C^1-conjugate to its linearization at the hyperbolic saddle point $(s, 0)$, and therefore solutions on the stable manifold converge exponentially to the fixed point. The decay rate is given by the eigenvalues of the Jacobian at the saddle point of system (5), hence we recover the result for $K = 0$ obtained in [22]. Observe that for $s < \bar{u}$ we obtain solitary waves of depression with the same qualitative properties.

Similarly to solitary waves, periodic traveling waves are obtained by choosing $h \in (h_c, h_s)$ in (8) where $h_c = F(u_c)$ and u_c is a solution of

$$F'(y) = 0 \quad \text{and} \quad \begin{cases} F''(y) > 0 & \text{when } y > \bar{u}, \\ F''(y) < 0 & \text{when } y < \bar{u}. \end{cases}$$

These periodic waves undulate about $u = u_c$.

2.2. Conditions for the existence of solitary traveling waves

We now derive algebraic conditions for the existence of homoclinic orbits, which give rise to solitary traveling wave solutions of (1) as we have just seen. At this point, our problem involves three interdependent parameters:

c ... the wave speed,
K ... the constant of integration,
s ... the level of the undisturbed water surface at infinity.

It turns out to be more convenient to eliminate the parameter K in favor of s using relation $F'(s)=0$ from the definition (10) of s. This leads us to the following change of parameters:

$$\Phi : (c, s) \longmapsto (c, K) = (c, \varphi(c, s)), \tag{11}$$

where

$$\varphi(c, s) = \frac{1}{14}s(-3s^3 + 2s^2 - 3s + c - 1). \tag{12}$$

Remark 2.2.

Notice that Φ is not bijective on R^2. For instance, for each $c \in R$ and K big enough the point (c,K) has no preimage. However, this happens precisely when there are no simple solutions of $F'(s)=0$, in which case system (6) does not have homoclinic orbits. In the remaining cases, observe that for each fixed $c \in R$ there exist two $s_1 \neq s_2$ such that $\varphi(c,s_1)=\varphi(c,s_2)=(c,K)$. This leads to a redundancy in the parameter regions, since for each c there are two values s which yield the same K, and hence the same F, which gives rise to the same phase portrait of system (6), and therefore also to the same wave solution. The difference between the two values of s is that one of them, say s_1, satisfies(10) and hence corresponds to the saddle point of system (6). The other value, s_2, which satisfies the reverse inequalities, gives rise to a center of the system. The solitary wave solution corresponding to the homoclinic orbit around this center $(s_2,0)$ decays to the undisturbed water level given precisely by the former value s_1, which means that this solitary wave solution is already obtained via the value s_1. To avoid this redundancy, we select the value s in (12) satisfying

$$\begin{cases} F''(s) < 0 & \text{when } s > \bar{u}, \\ F''(s) > 0 & \text{when } s < \bar{u}, \end{cases} \tag{13}$$

which makes the transformation (11) bijective on the relevant regions. In this way, we choose the value corresponding to the saddle point of the system, which is in accordance with the physical interpretation of s being the level of the undisturbed water surface at infinity.

Performing the substitution $K=\varphi(c,s)$ given in (12) facilitates our analysis considerably and we proceed to study conditions for the existence of homoclinic orbits in terms of the parameters s and c. The discussion in the proof of Proposition 2.1ensures that (6) has at most two critical points corresponding to the local extrema of F. It is easy to see that homoclinic orbits exist when there is at most one saddle point and one center point in the phase plane, cf. Fig. 2. This situation occurs when

(i)F has two distinct local extrema, and moreover,

(ii)both extrema lie either to the left or to the right of the invariant line $u = \bar{u}$.

To ensure condition (i) we study the roots of the discriminant of F', which in view of the substitution (12) is given by

$$\mathrm{Dis}(F'(u), u) = \alpha F''(s)^2 M(c, s), \tag{14}$$

where $\alpha<0$ is a real constant and $M(c,s)$ is a polynomial with no real roots, cf.Appendix A.1. Condition (ii) holds whenever $F'(\bar{u})$ is non-zero, i.e. we study the roots of

$$F'(\bar{u}) = \beta(s - \bar{u})N(c, s), \tag{15}$$

where $\beta>0$ is a real constant and $N(c,s)$ is a cubic polynomial in c and s. The algebraic curves corresponding to the zeros of these factors intersect precisely in one point (c^*,s^*) in the parameter plane, cf. Appendix A.1 and Fig. 3. We distinguish the following six regions:

$s > \bar{u}$	$s < \bar{u}$
R_1: $F''(s) < 0$ and $N(c,s) > 0$	R_4: $F''(s) > 0$ and $N(c,s) < 0$
R_2: $F''(s) > 0$ and $N(c,s) > 0$	R_5: $F''(s) < 0$ and $N(c,s) < 0$
R_3: $F''(s) > 0$ and $N(c,s) < 0$	R_6: $F''(s) < 0$ and $N(c,s) > 0$

(16)

Let us focus first on the regions where $s > \bar{u}$. Choosing (c,s) in R_3, we have that $F'(\bar{u}) < 0$ in view of (15) and (16), meaning that there is one extremum of F on each side of the invariant line $\bar{u}$. In this case, both extrema of F yield a center for system (6)which impedes the existence of a homoclinic connection. Hence, region R_3 gives rise to periodic orbits only (see Section 2.4). Choosing (c,s) in R_2 violates the condition on $F''(s)$ in (13), and hence we refrain from any further analysis (recall the redundancy discussed in Remark 2.2). Finally, (c,s) in R_1 yields F with two distinct local extrema to the right of $\bar{u}$ since $F'(\bar{u}) > 0$, and hence F has a local maximum in $s$in view of $F''(s)<0$. The homoclinic orbit based in the corresponding saddle point$(s,0)$ gives rise to a solitary traveling wave solution of (1) which propagates at speed$c$ and decays to the undisturbed water level s at infinity. Notice that all solitary wave solutions obtained in this way from parameters in region R_1 are waves of elevation. Applying the same reasoning to the regions where $s < \bar{u}$, we find that R_6 gives rise to periodic orbits only, whereas choosing parameters in region R_4 yield solitary waves of depression. We conclude that there are two algebraic curves A_1 and A_2 bounding the region $A=R_1 \cup R_4$ which admits solitary traveling wave solutions of (1):

$$\begin{cases} A_1 := \{F''(s) = -1 - 6s + 6s^2 - 12s^3 + c = 0\} \\ A_2 := \left\{s - \bar{u} = s + \dfrac{1+c}{14} = 0\right\}. \end{cases} \tag{17}$$

We summarize our conclusions in the following proposition.

Proposition 2.3.

Solitary wave solutions of (1) propagating at speed c *and decaying to the undisturbed water level* s *at infinity exist if and only if* $(c,s) \in A = R_1 \cup R_4$, *the region which is bounded by the algebraic curves* A_1 *and* A_2 *defined in (17) above. Parameters in* R_1 *yield solitary waves of elevation, whereas* R_4 *gives rise to solitary waves of depression.*

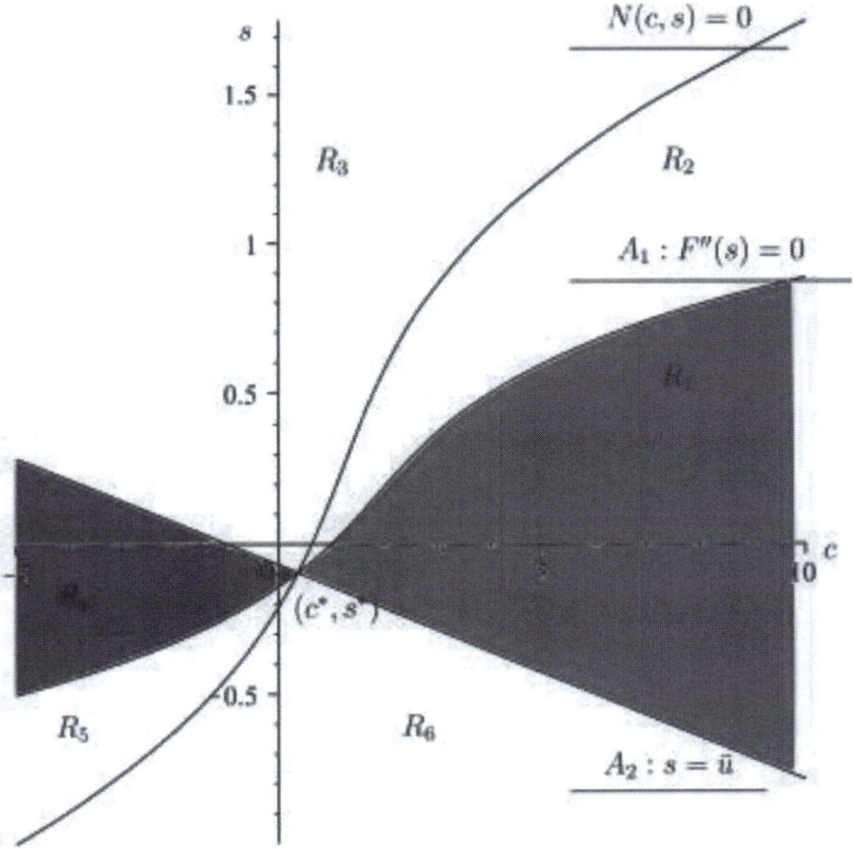

Figure 3. The shaded region $A = R_1 \cup R_4$ in the parameter plane (c,s) yields homoclinic orbits which give rise to solitary wave solutions of (3) traveling at speed c and decaying to the undisturbed water level s at infinity. When $s > \bar{u}$ we obtain solitary waves of elevation (R_1), whereas for $s < \bar{u}$ we obtain solitary waves of depression (R_4).

2.3. Solitary waves with compact support

It turns out that Eq. (1) admits weak solitary wave solutions in C^1 with compact support on R. This is essentially due to the fact that the planar system (5) is discontinuous along the straight line $u = \bar{u}$ (for a detailed account on various evolution equations arising in the context of nonlinear water waves which yield so-called "singular nonlinear traveling wave systems" we refer to [30]). The intuition behind this surprising observation is that the homoclinic orbit corresponding to the compactly supported solitary wave has finite existence time when the local maximum of F lies at the invariant line $\bar{u}$, cf. Fig. 4. This situation occurs precisely on the curve A_2, where $s = \bar{u}$, in which case the level line of the Hamiltonian corresponding to the homoclinic orbit based in the saddle point is $h_s = F(\bar{u})$ and $F'(\bar{u}) = 0$. Therefore, relation (8) simplifies to

$$v = \pm\sqrt{2\frac{F(\bar{u}) - F(u)}{u - \bar{u}}} = \pm\sqrt{(u - \bar{u})\, p(u)}, \tag{18}$$

where

$$p(u) := -F''(\bar{u}) - \frac{2}{3!}F^{(3)}(\bar{u})(u - \bar{u}) - \cdots - \frac{2}{5!}F^{(5)}(\bar{u})(u - \bar{u})^3.$$

In particular, the existence time of these homoclinic orbits is finite. Indeed, notice that

$$T(u, u_0) := \int_{u_0}^{u} \frac{dr}{\sqrt{(r - \bar{u})p(r)}}$$

is an elliptic integral and therefore finite, since $p(r)$ is a third degree polynomial with no repeated roots and $\bar{u}$ is not a root of $p(r)$. In view of (18) this yields

$$T(u(\xi), u_0) = \int_{u(\xi_0)}^{u(\xi)} \frac{dr}{\sqrt{(r - \bar{u})p(r)}} = \int_{\xi_0}^{\xi} \frac{\sqrt{(u - \bar{u})p(u)}}{\sqrt{(u - \bar{u})p(u)}} d\xi = \xi - \xi_0,$$

for a solution of $\dot{u}(\xi) = \sqrt{(u-\bar{u})p(u)}$ with initial data $u(\xi_0)=u_0$. Hence, the time it takes an orbit to get from $\bar{u}$ to m, where m is the non-trivial solution of $F(\bar{u}) = F(m)$, is given by

$$T := T(u(\xi), \bar{u}) - T(u(\xi), m) = \int_{\bar{u}}^{m} \frac{dr}{\sqrt{(r-\bar{u})p(r)}} < \infty.$$

By symmetry it follows that the solitary traveling wave solution corresponding to the orbit with $h_s = F(\bar{u})$ is defined on the finite interval $(-T,T)$. We may extend this solution to the real line by setting $u(\xi) = \bar{u}$ for $\xi \in R\setminus(-T,T)$. This is possible since $u = \bar{u}$ is a constant solution of (4) when $s = \bar{u}$. Furthermore, $u(\xi) \to \bar{u}$ as $\xi \to \pm T$ and therefore $v \to 0$ in view of (18), but $\dot{v} \to -1/2F''(\bar{u})$. This proves that the extension is C^1 with compact support in R. Therefore, $u \in H^1(R)$ is a weak solitary wave solution of (1) with compact support, see Remark 2.4. Notice that, when ξ approaches $\pm T$, the solution decays like

$$u(\xi) = \bar{u} - \frac{1}{4}F''(\bar{u})(\xi \pm T)^2 + O((\xi \pm T)^3) \tag{19}$$

which is readily checked.

Remark 2.4.

Eq. (3) may be rewritten as

$$(1-c)u + 3u^2 - 2u^3 + 3u^4 - 7\left(\frac{d}{d\xi}u\right)^2 = -\frac{1}{28}\frac{d^2}{d\xi^2}\left((1+c+14u)^2\right) - 14K. \tag{20}$$

We say that $u \in H^1_{\mathrm{loc}}(\mathbb{R})$ is a weak traveling wave solution of (1) if u satisfies (20) in the sense of distributions. Observe that the weak formulation of Eq. (1) reads

$$u_t + \partial_x\Big(-u - 7u^2 + (1-\partial_x^2)^{-1}(2u + 10u^2 - 2u^3 + 3u^4 - 7u_x^2)\Big) = 0,$$

which is equivalent to (20) for periodic traveling waves, or traveling waves which decay at infinity.

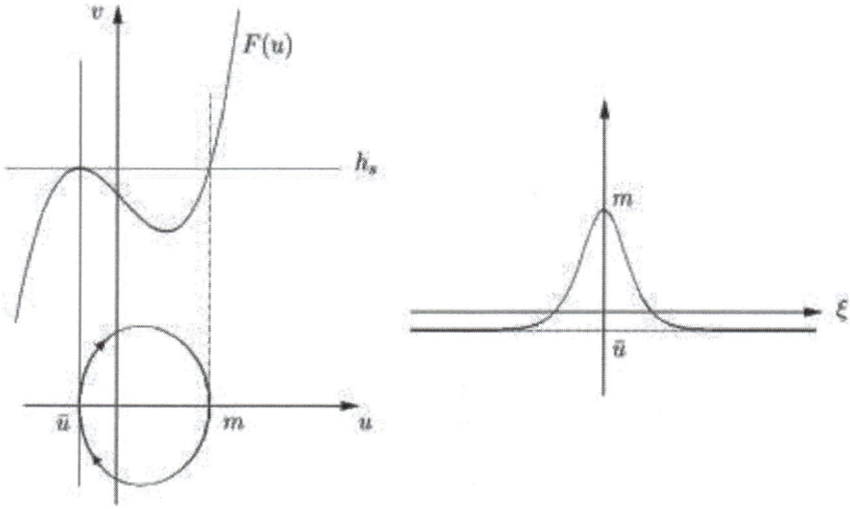

Figure 4. The choice of parameters $c \in R$ and $s = \bar{u}$ in F yields a homoclinic orbit with finite existence time, which gives rise to a solitary wave solution with compact support.

2.4. Peaked periodic waves

For parameters in the regions R_3 and R_6, cf. Fig. 3, the invariant line $\bar{u}$ lies between the two critical points of the polynomial F. In this case, the extrema of F yield two centers in the phase portrait of (6) which impedes the existence of solitary waves (and hence, the parameter s no longer accommodates the physical interpretation of the undisturbed water level at infinity). However, we show by continuous extension that there exist peaked periodic waves above and below the line $\bar{u}$, and periodic waves undulating about $\bar{u}$.

Indeed, for every $(c,s) \in R_3 \cup R_6$, periodic waves are obtained as in Proposition 2.1 by choosing $h_p \in (h_1, h_2)$, where $h_i = F(u_i)$ for $i = 1, 2$, and u_i is a solution of

$$F'(y) = 0 \quad \text{and} \quad \begin{cases} F''(y) > 0 & \text{when } y > \bar{u}, \\ F''(y) < 0 & \text{when } y < \bar{u}, \end{cases}$$

and employing (8). We will now treat the special case $h_p = F(\bar{u})$. Notice that, by construction,

$$h_p - F(u) = (u - \bar{u})(u - m_1)(u - m_2)q(u),$$

where $q(u)$ is a second order polynomial with no real roots and $m_i \neq \bar{u},\ i = 1, 2$, are the other two intersections of the horizontal line h_p with $F(u)$. Using (8), we obtain two heteroclinic orbits of the system (6) leaving and returning to the invariant line $u = \bar{u}$ given in terms of

$$v_i = \pm\sqrt{2(u - m_1)(u - m_2)q(u)}, \tag{21}$$

for $u \in (m_1, \bar{u})$ and $u \in (\bar{u}, m_2)$ respectively, which intersect the horizontal axis at m_1 and m_2, where $m_1 < \bar{u} < m_2$. Observe that for the topologically equivalent system (5), the existence times of these orbits are again finite and given in terms of

$$T_1 = \int_{m_1}^{\bar{u}} \frac{du}{\sqrt{2(u - m_1)(u - m_2)q(u)}} < \infty$$

and

$$T_2 = \int_{\bar{u}}^{m_2} \frac{du}{\sqrt{2(u - m_1)(u - m_2)q(u)}} < \infty.$$

Other than in Section 2.3 it is not possible to continuously extend the corresponding solutions by $\bar{u}$ on the real line to obtain solitary waves with compact support, since here $\bar{u}$ (or any other constant) does not satisfy Eq. (4). However, we may continue the solutions periodically and obtain peaked periodic waves u_i, cf. Fig. 5. The period of these waves is $2T_i$, and they have countably many points of discontinuity at the wave crests or troughs, $\xi=(2k+1)T_i$ where $k\in Z, i=1,2$, respectively. When $T_1=T_2$, we obtain C^2-periodic traveling waves u_P with period $4T_1$, undulating about the flat surface at $\bar{u}$. Indeed, a continuity argument guarantees the existence of parameters$(c,s)\in R_3\cup R_6$ such that $T_1=T_2$. In this case, the peaked periodic solutions $u_i$obtained from (21) may be glued together at $\xi=(2k+1)T_1, k\in Z$, cf. Fig. 5(c), which yields a smooth periodic solution u_P undulating about $\bar{u}$ defined on R by

$$u_P(\xi) = \begin{cases} u_1(\xi), & \text{when } \xi \in \bigcup\limits_{k=2m+1, m\in\mathbb{Z}} [(2k-1)T_1, (2k+1)T_1], \\ u_2(\xi), & \text{when } \xi \in \bigcup\limits_{k=2m, m\in\mathbb{Z}} [(2k-1)T_1, (2k+1)T_1]. \end{cases}$$

This continuation is C^2 since $u_1(\xi) \to \bar{u}$ as $\xi \nearrow (2k+1)T_1$, and therefore

$$\dot{u}_1 = v_1 = \pm\sqrt{2\frac{F(\bar{u})-F(u)}{u-\bar{u}}} \to \sqrt{-2F'(\bar{u})} \quad \text{and} \quad \ddot{u}_1 \to -\frac{F''(\bar{u})}{2} \quad \text{as } \xi \nearrow (2k+1)T_1,$$

and similarly for the limits as $\xi \searrow (2k+1)T_1$. The same reasoning shows that the continuation is C^2 at the lower bounds of the existence intervals.

3. PROPERTIES OF SOLITARY TRAVELING WAVES

The analysis in Section 2 shows that traveling wave solutions of (1) are symmetric with respect to the crest point, and that solitary waves tend (exponentially) to a constant on either side of their unique maximum or minimum. In the present section we will explore further properties regarding the shape of traveling waves. We determine how the wave amplitude, which is the positive difference between crest and trough, changes with respect to the wave speed. Furthermore, we prove that traveling waves are strictly monotone between crest and trough.

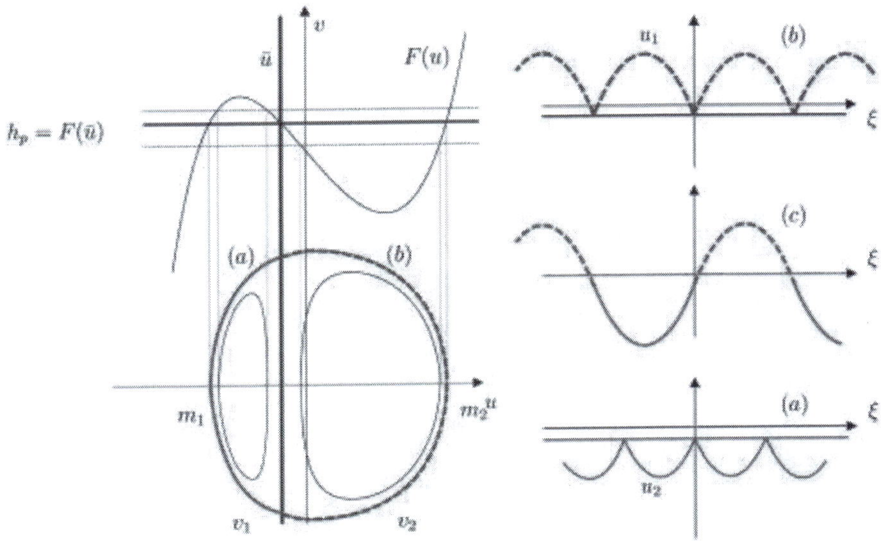

Figure 5. The choice of parameters $(c,s) \in R_3 \cup R_6$ gives rise to peaked periodic waves (a) and (b), as well as smooth periodic waves undulating about $\bar{u}$ (c).

3.1. Dependence of the amplitude on the wave speed

Our starting point is an algebraic expression for the change of $a = m - s$ with respect to c.

Lemma 3.1.

Let F *be the polynomial defined in* (9) *with a fixed* K *and let* (s,m) *be a solution of*

$$\begin{cases} F'(s) = 0, \\ F(s) - F(m) = 0, \end{cases} \tag{22}$$

where $s \neq \bar{u}$*. Then, for* $a = m - s \in R$*, we have that*

$$\partial_c a = \frac{-1/28}{F'(m)F''(s)}\Big((s^2 - m^2)F''(s) + 2s\,F'(m)\Big). \tag{23}$$

In the following proposition, we study the sign of (23) to determine the regions in the parameter set A which give rise to solitary waves whose amplitude |a| is increasing or decreasing with respect to the wave speed c, cf. Fig. 6.

Proposition 3.2.

We distinguish between the following cases:

- *For* $(c,s)\in R_1$ *we obtain solitary waves of* elevation *whose amplitude is strictly* increasing *with* c *in region* (I), *and* decreasing *with* c *in regions* (II) *and* (III), *for fixed* K.
- *For* $(c,s)\in R_4$ *we obtain solitary waves of* depression *whose amplitude is strictly* decreasing *with* c *in region* (IV) *and* increasing *with* c *in region* (V), *for fixed* K.

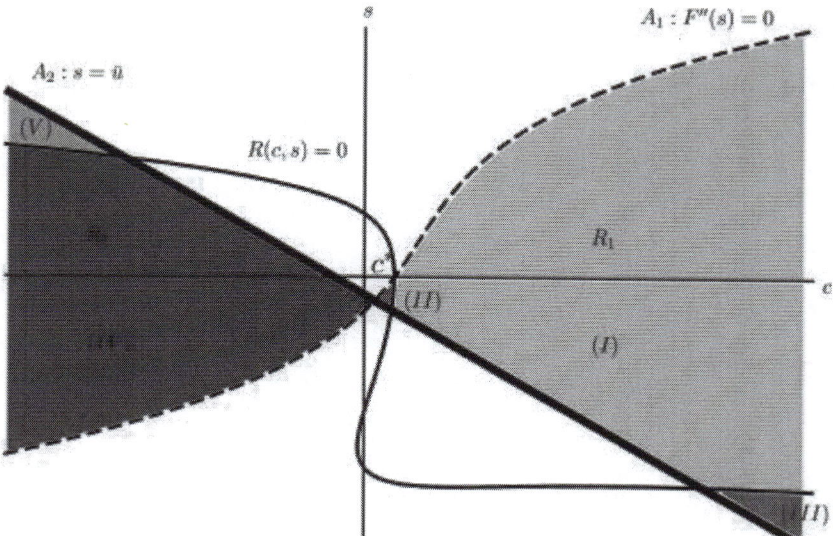

Figure 6. Choosing parameters in the lighter shaded regions (I) and (V) we obtain solitary waves which *increase* with the wave speed c, whereas solutions corresponding to parameters in the darker shaded regions (II), (III) *and* (IV) *decrease* with respect to c.

Proof of Proposition 3.2.

Denote $a=m-s$, so that the wave amplitude is given by $|a|$, and $a<0$ for waves of depression (in R_4) whereas $a>0$ for waves of elevation (in R_1). Observe that $F'(m)>0$ for all $(c,s)\in A$ and recall that $F''(s)<0$ in R_1 and $F''(s)>0$ in R_4. Therefore, and in view of (23), it suffices to study the sign of $(s^2-m^2)F''(s)+2sF'(m)$,

which in view of (11) yields

$$\frac{1}{14}(s-m)^2\underbrace{(6m^2s-4sm+12ms^2-2s^2+6s^3-1+c)}_{=:Q_{c,s}(m)}. \tag{24}$$

Recall that (s,m) solves (22) which reads

$$\frac{1}{140}(s-m)^2P_{c,s}(m) \tag{25}$$

where

$$P_{c,s}(m):=(6m^3-5m^2+12m^2s+10m+18ms^2-10sm+5-5c-15s^2+24s^3+20s).$$

.

Since we are interested in solutions $s\neq m$, we study the system

$$\begin{cases}Q_{c,s}(m)=0\\P_{c,s}(m)=0\end{cases} \tag{26}$$

which has a solution if and only if $Q_{c,s}(m)$ and $P_{c,s}(m)$ have a common root, i.e. if their resultant with respect to m,

$$\begin{aligned} R(c,s) &:= Res(Q_{c,s}(m), P_{c,s}(m), m) \\ &= 31104s^9 - 10368s^8 + 32832s^7 + (-15552c + 39456)s^6 \\ &\quad + (-3816 - 864c)s^5 + (23472 - 3312c)s^4 + 24(c-1)(108c - 593)s^3 \\ &\quad + 24(33c + 107)(c-1)s^2 - 690(c-1)^2 s + 36(c-1)^3, \end{aligned} \tag{27}$$

is zero (see Appendix A.2 for a discussion on the involved curves). Hence, system (26)has a solution, i.e. $\partial_c a = 0$, only along the algebraic curve $R(c,s)=0$, meaning that within the regions in A separated by this curve, the sign of $\partial_c a$ is constant. Hence, it suffices to pick one pair of parameters $(\bar{c}, \bar{s})$ in each of these regions and compute the values of m and Q to determine the sign of $\partial_c a$ in view of expression (23). For example, let $\bar{s}_1 = -0.1$ and $\bar{c}_1 = 1.5$ in R_1 then, computing the corresponding m_1, we find that $Q_{\bar{c}_1,\bar{s}_1}(m_1) > 0$. Therefore, since $F'(m)<0$ and $F''(s)<0$ in R_1, we obtain that $\partial_c a > 0$. Hence, the amplitude $|a|$ of solitary wave solutions of (1) arising from parameters in the region denoted by (I) in Fig. 6 is increasing with respect to the wave speed c. To provide an example for waves of depression, pick $\bar{s}_4 = -0.5$ and $\bar{c}_4 = 15$ in R_4, which yields m_4 such that $Q_{\bar{c}_4,\bar{s}_4}(m_4) < 0$. Therefore, since $F'(m)<0$ and $F''(s)>0$ in R_4, we obtain that $\partial_c a > 0$. In view of the fact that $a<0$ in R_4 this means that the amplitude $|a|$ of solitary waves with parameters in region (IV) is decreasing with respect to c. The results for the other regions (II), (III) and (V) can be obtained in exactly the same way. □

Proof of Lemma 3.1.

Consider F in (9), regarding it as a polynomial in u and c, and define

$$f(u,c) := F(u) = Ku + \frac{1-c}{28}u^2 + \frac{1}{14}u^3 - \frac{1}{28}u^4 + \frac{3}{70}u^5.$$

Then the first equation in (22) rewrites as

$$f_u(s,c) = 0,$$

where subscripts denote partial differentiation. Implicit differentiation with respect to c of the last equation yields

$$f_{uu}(s,c)\dot{s} + f_{uc}(s,c) = F''(s)\dot{s} - \frac{1}{14}s = 0,$$

where . denotes differentiation with respect to c, and therefore

$$\dot{s} = \frac{1}{14}\frac{s}{F''(s)}.$$

The second equation in (22) reads

$$f(s,c) - f(m,c) = 0,$$

which upon implicit differentiation yields

$$f_u(s,c)\dot{s} + f_c(s,c) - f_u(m,c)\dot{m} - f_c(m,c) = 0$$

so

$$\dot{m} = \frac{f_c(s,c) - f_c(m,c)}{f_u(m,c)} = \frac{-1/28}{F'(m)}(s^2 - m^2).$$

Since $\dot{a} = \dot{m} - \dot{s}$, this proves (23). □

3.2. Monotonicity

We show that the profile u of a solitary traveling wave solution of (1) is monotone from the undisturbed water level s to its maximum or minimum m, with precisely one inflection point on either side of the wave crest or trough. To this end, consider the right hand side of(8),

$$v = \sqrt{\underbrace{2\frac{F(s) - F(u)}{u - \bar{u}}}_{=:D(u)}}.$$

We claim that, as a function of u, this expression has a unique critical point between s and m. Since the square root is monotonous, it suffices to consider the number of critical points of the discriminant $D(u)$ of this expression. By construction, $D(u)$ has a critical point at s. We will show that there exists precisely one critical point in (s,m), which corresponds to the unique inflection point of the wave profile u between its trough and crest. To this end, we study $D'(u)$ and prove that it has exactly one real root to the right of s. For simplicity, we will give the proof only in the case $u > \bar{u}$, the other case can be proven in exactly the same way. Indeed, consider the numerator of $D'(u)$, which in view of (11) yields

$$(s - u)I_{c,s}(u), \tag{28}$$

where $I_{c,s}(u)$ is a fourth-order polynomial in u whose coefficients depend polynomially on the parameters s and c. In Appendix A.3 we prove that the number of roots of $I_{c,s}(u)$ in (s,∞) remains constant if we vary the parameters $(c,s) \in R_1$ (cf. also Lemma 3.6 in [31]). Therefore, it suffices to evaluate the polynomial at any point $(\bar{c}, \bar{s}) \in R_1$ and to deduce that the resulting univariate polynomial has a unique real root in $(\bar{s}, \infty)$ employing Sturm's method (cf. [32]). We conclude that $D'(u)$ has a unique real root to the right of s for parameters in R_1. The same reasoning shows that this holds also for $u < \bar{u}$ with parameters in R_2, and for compactly supported solitary waves with parameters on $s = \bar{u}$, cf. Appendix A.3. This proves the claim.

4. TRAVELING WAVES OF THE CAMASSA–HOLM EQUATION

We would like to point out that the method to prove existence of traveling waves put forward in Section 2 is applicable to a wide class of nonlinear dispersive evolution equations. As an example, we apply our approach to the Camassa–Holm equation, which is usually written in the form

$$u_t + 2\kappa\, u_x - u_{txx} + 3uu_x = 2u_x u_{xx} + uu_{xxx}, \quad (29)$$

for $x \in R, t > 0$ and $\kappa \in R$. For traveling waves $u(x, t) = u(x - ct)$, Eq. (29) takes the form

$$u''(u - c) + \frac{(u')^2}{2} + K + (c - 2\kappa)u - \frac{3}{2}u^2 = 0,$$

where K is a constant of integration. If instead of u we study the translate

$$w = u - c,$$

the previous equation reads

$$w''\, w + \frac{1}{2}(w')^2 + F'(w) = 0, \quad (30)$$

where

$$F(w) = A\,w - B\,w^2 - \frac{1}{2}w^3, \quad (31)$$

with constants $A = K - 2\kappa c - \frac{1}{2}c^2$ and $B = c + \kappa$. Now, Eq. (30) is of the form (4) and we may prove existence of traveling wave solutions as in Section 2:

Theorem 4.1.

There exist solitary and periodic traveling wave solutions of the Camassa–Holm equation (29) *for every* c, K *and* κ *satisfying*

$$-\frac{2}{3}B^2 < A < -\frac{1}{2}B^2, \tag{32}$$

with constants A *and* B *defined as above. All solitary waves are symmetric with respect to their unique maximum/minimum and tend exponentially to a constant at infinity. Periodic waves exist also for* $-\frac{1}{2}B^2 < A < 0$.

Proof.

Bounded orbits in the phase plane associated to (30) give rise to traveling waves of (29)in view of Proposition 2.1 as before. To work out conditions for the existence of homoclinic and periodic orbits we proceed along the lines of the proof of Proposition 2.3. Bounded orbits exist as long as the local extrema of F are distinct, i.e. when the discriminant of F', discrim$(F',w)=6A+4B^2$, is greater than zero. This yields the lower bound in (32). To guarantee the existence of homoclinic orbits we have to ensure that there is one saddle point and one center point in the phase plane. To this end, we study the relation $F(s)=F(0)$, where s is the solution of $F'(s)=0$ with $F''(s)<0$ for $s>0$ and $F''(s)>0$ for $s<0$, which yields the curve $A=-\frac{1}{2}B^2$ marking the upper bound in (32). □

Proposition 4.2.

There exist peaked continuous solitary traveling wave solutions of the Camassa–Holm equation (29) *for* c, K *and* κ *satisfying* $A=-\frac{1}{2}B^2$.

Proof.

For parameters satisfying $A=-\frac{1}{2}B^2$ we obtain homoclinic orbits which give rise to continuous solitary traveling wave solutions of (29) with a peaked crest. Indeed, for this choice of parameters $h_s=F(s)=0$, so the relation equivalent to (8) reduces to

$$w'=\pm\sqrt{\frac{-2F(w)}{w}}=\pm|B+w| \quad \text{and} \quad w''=\pm 1.$$

Hence, $w'\to\pm|B|$ when $w\to 0$, so there is a discontinuity of w' at $(0,\pm|B|)$, the crestpoint of the solitary wave solution.

However, it is straightforward to check that such a solution still satisfies the Eq. (30) in this point. Thus, we obtain solitary wave solutions with a peaked crest.

Remark 4.3.

Solitary traveling waves decaying to the flat surface at zero are known to exist when $c > 2\kappa$, cf. [5]. This result is reflected in our condition $A > -\frac{2}{3}B^2$ for $K=0$. Moreover, we recover the fact that peaked solitons (peakons) exist when $\kappa=0$, cf. [33], from the relation $A = -\frac{1}{2}B^2$ for $K=0$. The peakons and their periodic counterparts are orbitally stable, cf. [34], [35] and [36]. For a discussion on periodic traveling waves of the Camassa–Holm equation we refer to [21] and [14].

ACKNOWLEDGMENTS

The authors wish to thank Jordi Villadelprat for helpful comments and stimulating discussions regarding some properties of solitary traveling wave solutions. The first author is partially supported by a MCYT-FEDER grant number MTM2008-03437 and by a CIRIT grant number 2009SGR 410. The second author is supported by the FWF project J3452 "Dynamical Systems Methods in Hydrodynamics" of the Austrian Science Fund.

APPENDIX. ALGEBRAIC CURVES

We want to provide some remarks on the algebraic curves involved in our analysis and show that our figures display correctly the graphs of the corresponding expressions. We only exemplify the procedure by proving some selected cases. First, recall a result on the number of roots of polynomials (cf. [31]), which we will use repeatedly:

Lemma A.1.

Set $\Omega = \mathbb{R}$ *and consider a family of real polynomials whose coefficients depend continuously on a real parameter* b,

$$G_b(x) = g_n(b)x^n + g_{n-1}(b)x^{n-1} + \cdots + g_1(b)x + g_0(b).$$

Suppose there exists an open interval $I \subset \mathbb{R}$ *such that:*

(i)*There is some* $b_0 \in I$*, such that* $G_{b_0}(x)$ *has exactly* k *simple roots on* Ω.

(ii)*For all* $b \in I$*, the discriminant of* G_b *with respect to* x *is different from zero.*

(iii)*For all* $b \in I, g_n(b) \neq 0$.

Then for all $b \in I, G_b(x)$ *has exactly* k *simple roots on* Ω.

Moreover, if $\Omega = \Omega_b := (c(b), \infty) \subset \mathbb{R}$ *for some continuous function* $c(b)$ *the same result holds if we add the hypothesis:*

(iv) *For all* $b \in I, G_b(c(b)) \neq 0$.

The intuition behind this result is as follows: In view of the hypotheses (i)–(iv), the roots of $G_b(x)$ depend continuously on b. Assumptions (iii) and (iv) impede possible bifurcations of roots from infinity or from the boundary of Ω when varying $b \in I$. Moreover, assumption (ii) prevents the appearance of multiple real roots in the interior of Ω. Therefore, the number of roots of $G_b(x)$ is constant in Ω when the parameter $b \in I$ varies, and hence $G_b(x)$ has k simple roots for all $b \in I$ in view of assumption (i).

A.1. On the curves in Section 2.2

We want to ensure that the algebraic curve $M(c,s)=0$ in (14) has no real roots, and that $N(c,s)$ in (15) has a unique root for every choice of parameters (c,s). Furthermore, we want to show that the curves $N(c,s)=0$ and A_1, A_2 of (17) all intersect in precisely one point. We have

$$M(c,s) = 243c^2 + \left(-900s - 778 + 540s^2 - 1080s^3\right)c$$
$$+823 + 1284s + 480s^2 + 480s^3 + 2700s^4 - 1296s^5 + 1296s^6,$$

which we regard as a polynomial in c with parameter s. To show that it has no real roots, we check that the conditions of Lemma A.1 are satisfied for $k=0$. Assumption (iii) holds, since the coefficient of the highest order term is constant. Computing the discriminant of $M(c,s)$ with respect to c yields

$$-16\left(18s^2-6s+23\right)^3,$$

for which it is straightforward to prove that it has no zeros in R. To check assumption (i) we choose, for example, $s=-1$ which gives

$$M(c,-1)=243c^2+1742c+4831>0.$$

In view of Lemma A.1, we find that $M(c,s)$ is strictly positive for all $c,s\in R$. Next we focus on

$$N(c,s)=3c^3+(-42s+37)c^2+\left(-476s+588s^2+3397\right)c+6076s^2-8232s^3-8666s-2125.$$

It is straightforward to show that assumptions (ii) and (iii) hold. Regarding (i), we choose for example $s=1$ and find that

$$N(c,1)=(3c-11)(c^2+2c+1177)$$

which clearly has a unique root. Hence, for each $s\in R$ there exists precisely one $c\in R$ such that $N(c,s)=0$. To prove that the curves A_1,A_2 and $N(c,s)=0$ all intersect in precisely one point, observe that the A_i are linear in c and it is therefore easy to check that they intersect at a point (c^*,s^*) where s^* is the root of the cubic polynomial $P^*:=-2-20s+6s^2-12s^3$ and c^*, the corresponding value on the curve A_2, is given by the unique real solution of $3c^3+30c^2+1031c-368=0$ as stated in Theorem 1.1. To see that $N(c,s)=0$ also intersects in that point, compute the resultant of A_2 and N with respect to c and find that the resulting polynomial is just a factor of P^* which proves the claim.

A.2. On the curves in Section 3.1

In this subsection, we want to discuss the curves in (24), (25) and (27). We show that the curve $R(c,s)=0$ intersects the curve A_2 three times whereas it intersects A_1 only once, which yields the different regions depicted in Fig. 6. To prove the latter result, we solve both A_1 and A_2 for c and plug the resulting expressions into $R(c,s)$. We obtain two univariate polynomials in s for which it is straightforward to show that they have three roots (two negative and one positive) and one root (in zero), respectively. To show that the polynomial expressions defining the involved curves in (24), (25) and (27) yield a unique root for each $(c,s)\in\mathbb{R}^2$, we again employ Lemma A.1 and check that the assumptions are satisfied. We exemplify the procedure by showing that the result is true for the curve $P_{c,s}(m)=0$, where

$$P_{c,s}(m) = -6m^3 + (5-12s)m^2 + (-18s^2+10s-10)m - 24s^3 + 15s^2 - 20s - 5 + 5c.$$

Note that we are now dealing with a polynomial which depends on two parameters. Assumption (iii) holds in view of the fact that the coefficient of the highest order term is constant, and it is straightforward to check (i) choosing parameters $(\bar{c},\bar{s})$ which yield that $P_{\bar{c},\bar{s}}(m)$ has precisely one root. To prove assumption (ii), we compute the discriminant of $P_{c,s}(m)$ with respect to m and obtain

$$\begin{aligned}\mathrm{Dis}(P_{c,s}(m),m) &= -24300c^2 + (120600s - 75600s^2 + 73100 + 151200s^3)c\\ &\quad - 70300 - 96200s^2 - 163600s - 504000s^4 + 259200s^5 + 2400s^3 - 259200s^6.\end{aligned}$$

To show that $\mathrm{Discrim}(P_{c,s}(m))$ is different from zero, we repeat the scheme for this polynomial in c with parameter s and ensure again that the assumptions ofLemma A.1 hold with $k=0$.

A.3. On the curves in Section 3.2

The goal of this subsection is to prove that for the polynomial

$$\begin{aligned}I_{c,s}(u) &= -168u^4 + (90-15c-168s)u^3 + \left(10c - 130 - 168s^2 + 90s - 15sc\right)u^2\\ &\quad + (-50 + 10sc + 20c - 130s + 90s^2 - 168s^3 - 15s^2c)u\\ &\quad - 5 - 50s + 90s^3 + 5c^2 + 20sc - 130s^2 - 168s^4 + 10s^2c - 15s^3c,\end{aligned}$$

obtained in (28), the number of roots do not change if we vary the parameters (c,s) in the admissible region A. We will again employ Lemma A.1 above for $I=A$ and $\Omega_{c,s}=(s,\infty)$. Indeed, assumption (iii) holds since the highest coefficient of $I_{c,s}(u)$ is constant. To check that (iv) is satisfied, we evaluate the polynomial at $u=s$ and find that

$$I_{c,s}(s) = (s-\bar{u})F''(s).$$

These factors are exactly the relations which bound the admissible parameter region A, and hence they do not vanish in the interior of A. Note, however, that solitary waves with compact support arise from a choice of parameters (c,s) on the curve $A_2 = \{s - \bar{u} = 0\}$, so we need a separate argument in that case which will be carried out below. Next we study the discriminant of $I_{c,s}(u)$ with respect to u,

$$\text{Dis}(I_{c,s}(u), u) = D_1(c, s)D_2(c, s), \tag{A.1}$$

where the factors on the right hand side of the above expression are given below. We claim that the algebraic curves corresponding to the zeros of these factors lie outside of A. To see this, observe that the curves $A_1, A_2, \{D_1(c,s)=0\}$ and $\{D_2(c,s)=0\}$ intersect precisely once in the point (c^*,s^*). Then, we choose some $c_1<c^*$ and find that $D_1(c_1,s)>A_1(c_1,s)>A_2(c_1,s)>D_2(c_1,s)$, whereas for any $c_2>c^*$ we obtain the reverse order. Hence, the discriminant of $I_{c,s}(u)$ does not vanish in A, which proves the claim. Therefore, the assumptions (ii)–(iv) of Lemma A.1 hold, and we find that the number of roots of $I_{c,s}(u)$ is constant in the interior of A.

We now provide a separate but similar argument which asserts that this result holds also for parameters on the curve A_2. Indeed, for (c,s) on A_2, i.e. when $s = \bar{u}$, we find that

$$D'(u) = \tilde{I}_c(u),$$

where $\tilde{I}_c(u)$ is a cubic polynomial in u whose coefficients depend polynomially on c. Along the lines of the above proof we argue that $\tilde{I}_c(u)$ has a unique real root in $(\bar{u}, \infty)$. Indeed, no bifurcations of roots occur at infinity, and evaluating $\tilde{I}_c(u)$ at the boundary $u = \bar{u}$ yields a cubic polynomial in c which vanishes only in $c^* \notin A$. Using Sturm's method (cf. [32]) we show that the discriminant of $\tilde{I}_c(u)$ with respect to u does not vanish. Hence, the number of roots is constant, and choosing any $\bar{c}$ we find that $\tilde{I}_{\bar{c}}(u)$ has a unique real root in $(\bar{u}_{\bar{c}}, \infty)$.

We conclude with a discussion of the polynomials in expression (A.1),

$$D_1(c,s) = -32928s^3 + (1764c + 22344)s^2 - (84c^2 + 1148c + 28504)s + 3c^3 + 44c^2 + 7919c - 5842$$

And

$$\begin{aligned} D_2(c,s) = {} & 30375c^5 + (67500s^2 - 135000s^3 + 93100 + 567900s)c^4 \\ & + (2083200s^2 - 3518100s^4 + 408900s + 162000s^6 + 1280400s^3 - 162000s^5 + 880703)c^3 \\ & + (-368064s - 4730400s^6 - 8347536s^2 + 5443200s^7 + 6777000s^4 - 11014128s^3 \\ & - 25691040s^5 - 3605574)c^2 + (-44997120s^7 + 4400084 - 15110352s + 23678784s^5 \\ & - 9163584s^6 + 60963840s^8 - 22971024s^2 - 175259552s^3 - 58261680s^4)c + 227598336s^9 \\ & - 138184704s^8 + 31667136s + 301625856s^7 + 71568192s^3 + 152350848s^6 + 1062232 \\ & + 54393984s^5 + 49720800s^2 + 187454304s^4. \end{aligned}$$

We will only discuss the latter curve and employ Lemma A.1 again for $I \times \Omega = A$. Note that assumption (iii) holds in view of the fact that the coefficient of the highest order term is constant. Computing the discriminant with respect to c yields

$$\begin{aligned} \mathrm{Dis}(D_2, c) = {} & \alpha_0(11664000s^{12} - 23328000s^{11} + 367804800s^{10} - 487728000s^9 \\ & + 3390049800s^8 - 2253805200s^7 + 4960871884s^6 + 2160459976s^5 \\ & + 1280057526s^4 + 4059678628s^3 + 1729573411s^2 + 1328288220s + 695918709) \\ & \times (84672s^4 + 22512s^3 + 76402s^2 + 58822s + 16767)^3 \\ & \times \underbrace{(16767 + 29606s - 12083s^2 - 4040s^3 + 20160s^4)}_{=:\tilde{D}(s)}{}^2, \end{aligned}$$

where $\alpha_0 > 0$ is a real constant. It is straightforward to see that the first two factors of the above expression have no real roots, whereas the last factor $\tilde{D}(s)$ vanishes for two values s_1 and s_2, meaning that $D_2(c, s_i)$ may have multiple roots in that case. To ensure that assumption (ii) holds, we have to prove that these values do not lie in A. To this end, we use Sturm's method to derive rational upper and lower bounds for s_i such that $s_1 \in [\underline{s}_1, \bar{s}_1]$ and $s_2 \in [\underline{s}_2, \bar{s}_2]$, to bound the curves A_i from above and below by rational constants. For example, we find that $s_1 \in [-\frac{131}{128}, -\frac{65}{64}] =: I_1$. We claim that all values of D_2 in the strip defined by I_1 lie outside of A. To this end we compute a bound for A_2,

$$M_1 := \min_{s \in I_1}\{c \in \mathbb{R} : A(c, s) = 0\} = \frac{423}{32} \in \mathbb{Q},$$

and construct a rational univariate polynomial

$$\begin{aligned}\tilde{D}_2(c) &= 30375c^5 - \frac{8897979325}{32768}c^4 - \frac{602652229378808443}{274877906944}c^3 \\ &+ \frac{3196448247290763459}{137438953472}c^2 + \frac{26418684722558295308 63}{70368744177664}c - \frac{17460388511693021202337}{35184372088832},\end{aligned}$$

using the upper and lower bounds of the interval I_1 such that $D_2(c, s) < \tilde{D}_2(c)$. Then, it is fairly straightforward to see that $D_2(c, s) - M_1 < \tilde{D}_2(c) - M_1 < 0$, which proves the claim. Repeating this procedure with the root s_2 and the other bounding curves A_i shows that in the region A all involved curves are displayed correctly.

REFERENCES

1. R. Camassa, D.D. Holm, An integrable shallow water equation with peaked solitons, Phys. Rev. Lett. 71 (11) (1993) 1661–1664.

2. A. Boutet de Monvel, A. Kostenko, D. Shepelsky, G. Teschl, Long-time asymptotics for the Camassa–Holm equation, SIAM J. Math. Anal. 41 (4) (2009) 1559–1588.

3. A. Constantin, On the scattering problem for the Camassa–Holm equation, Proc. R. Soc. Lond. Ser. A 457 (2001) 953–970.

4. A. Constantin, V.S. Gerdjikov, R.I. Ivanov, Inverse scattering transform for the Camassa–Holm equation, Inverse Problems 22 (6) (2006) 2197–2207.

5. A. Constantin, W. Strauss, Stability of the Camassa–Holm solitons, J. Nonlinear Sci. 12 (4) (2002) 415–422.

6. K. El Dika, L. Molinet, Exponential decay of H1-localized solutions and stability of the train of N solitary waves for the Camassa–Holm equation, Philos.

7. Trans. R. Soc. Lond. Ser. A 365 (1858) (2007) 2313–2331.

8. A. Constantin, J. Escher, Wave breaking for nonlinear nonlocal shallow water equations, Acta Math. 181 (2) (1998) 229–243.

9. A. Bressan, A. Constantin, Global conservative solutions of the Camassa–Holm equation, Arch. Ration. Mech. Anal. 183 (2007) 215–239.

10. A. Bressan, A. Constantin, Global dissipative solutions of the Camassa–Holm equation, Anal. Appl. 5 (1) (2007) 1–27.

11. A. Constantin, H.P. McKean, A shallow water equation on the circle, Comm. Pure Appl. Math. LII (1999) 949–982.

12. F. Gesztesy, H. Holden, Algebro-geometric solutions of the Camassa–Holm hierachy, Rev. Mat. Iberoam. 19 (2003) 73–142.

13. J. Lenells, Traveling wave solutions of the Camassa–Holm equation, J. Differential Equations 217 (2) (2005) 393–430.

14. A. Constantin, Nonlinear Water Waves with Applications to Wave–Current Interactions and Tsunamis, CBMS-NSF R SIAM, Philadelphia, 2011.

15. R.S. Johnson, Camassa–Holm, Korteweg–de Vries and related models for water waves, J. Fluid Mech. 455 (2002) 63–82.

16. A. Constantin, D. Lannes, The hydrodynamical relevance of the Camassa–Holm and Degasperis–Procesi equations, Arch. Ration. Mech. Anal. 192 (2009) 165–186.

17. T. Kato, Quasi-linear equations of evolution, with applications to partial differential equations, Spectr. Theory Differential Equations 448 (1975) 25–70.

18. N. Duruk Mutlubas, On the Cauchy problem for a model equation for shallow water waves of moderate amplitude, Nonlinear Anal. RWA 14 (5) (2013) 2022–2026.

19. N. Duruk Mutlubas, Local well-posedness and wave breaking results for periodic solutions of a shallow water equation for

waves of moderate amplitude, Nonlinear Anal. TMA 97 (2014) 145–154.

20. N. Duruk Mutlubas, A. Geyer, B.-V. Matioc, Non-uniform continuity of the flow map for an evolution equation modeling shallow water waves of moderate amplitude, Nonlinear Anal. RWA 17 (2014) 322–331.

21. Y. Mi, C. Mu, On the solutions of a model equation for shallow water waves of moderate amplitude, J. Differential Equations 255 (8) (2013) 2101–2129.

22. A. Constantin, J. Escher, Well-posedness, global existence and blowup phenomena for a periodic quasi-linear hyperbolic equation, Comm. Pure Appl. Math. LI (1998) 475–504.

23. A. Geyer, Solitary traveling water waves of moderate amplitude, J. Nonlinear Math. Phys. 19 (supp 01) (2012) 1240010. p. 12.

24. N. Duruk Mutlubas, A. Geyer, Orbital stability of solitary waves of moderate amplitude in shallow water, J. Differential Equations 255 (2) (2013) 254–263.

25. M. Grillakis, J. Shatah, W. Strauss, Stability theory of solitary waves in the presence of symmetry I, J. Funct. Anal. 74 (1) (1987) 160–197.

26. F. Dumortier, J. Llibre, J.C. Artés, Qualitative Theory of Planar Differential Systems, Springer, Berlin, 2006.

27. J. Guggenheimer, P. Holmes, Nonlinear Oscillations, Dynamical Systems, and Bifurcations of Vector Fields, Springer, 1983.

28. L. Perko, Differential Equations and Dynamical Systems, Springer, New York, 2006.

29. G. Teschl, Ordinary Differential Equations and Dynamical Systems, in: Graduate Studies in Mathematics, vol. 140, AMS, Providence, RI, 2012.

30. J. Sotomayor, Lições de Equações Diferenciais Ordinárias, in: Projeto Euclides, vol. 11, Instituto de Matemática Pura e Aplicada, Rio de Janeiro, 1979.

31. J. Li, Singular Nonlinear Travelling Wave Equations: Bifurcations and Exact Solutions, in: Mathematics Monograph Series, vol. 27, Science Press, Beijing, 2013.

32. J.D. García-Saldaña, A. Gasull, H. Giacomini, Bifurcation values for a family of planar vector fields of degree five, DCDS-A (2014) in press.

33. J. Stoer, R. Bulirsch, Introduction to Numerical Analysis, Springer Verlag, New York Heidelberg, 1980.

34. R. Camassa, D.D. Holm, J.M. Hyman, A new integrable shallow water equation, Adv. Appl. Mech. 31 (31) (1994) 1–33.

35. A. Constantin, W. Strauss, Stability of peakons, Comm. Pure Appl. Math. 53 (2000) 603–610.

36. J. Lenells, Stability of periodic peakons, Int. Math. Res. Not. 10 (2004) 485–499.

37. J. Lenells, A variational approach to the stability of periodic peakons, J. Nonlinear Math. Phys. 11 (2) (2004) 151–163.

6

CHAPTER

Nonlinear Stability of Flock Solutions in Second-Order Swarming Models

J.A. Carrillo , Y. Huang , S. Martin

Department of Mathematics, Imperial College London, London, SW7 2AZ, U

ABSTRACT

In this paper we consider interacting particle systems which are frequently used to model collective behaviour in animal swarms and other applications. We study the stability of orientationally aligned formations called flock solutions, one of the typical patterns emerging from such dynamics. We provide an analysis showing that the nonlinear stability of flocks in second-order models entirely depends on the linear stability of the first-order aggregation equation. Flocks are shown to be nonlinearly stable as a *family* of states under reasonable assumptions on the interaction potential. Furthermore, it is tested numerically that commonly used potentials satisfy these hypotheses and the nonlinear stability of flocks is investigated by an extensive case study of uniform perturbations.

1. INTRODUCTION

Self-organization and pattern formation are ubiquitous in nature and science, ranging from animal aggregation [1], [2], [3], [4] and [5] and biological systems [6] and [7] to self-assembly of nano-particles [8] and [9]. The intense studies during the last two decades are on both individual-based systems and continuum equations [10], [11], [12] and [13]. One of the essential features in these models is the non-zero characteristic speed of the individual agents, which has been modelled in different ways. The speed can be assumed to be constant with a direction based on the averages of the neighbours [14] or to be driven by random noise [15]. On the other hand, a large class of models consist of self-propelled particles powered by biological or chemical mechanisms with friction forces, resulting in a preferred characteristic speed.

In general, the particles with non-zero equilibrium speed do not form any recognizable patterns [4] and [16], and interactions within the group have to be included to generate interesting spatial configurations. Most of these interaction forces have been taken into account in the combination of three effects: alignment, repulsion, and attraction; also called the "first principles of swarming". The basic mechanisms account for collisional avoidance and comfort regions (repulsion), grouping and socialization (attraction), and mimetic synchronization (alignment). The combination of these three effects goes back to fish school modelling [17] and [18] and these basic ideas have been improved and applied to several animal species including different mechanisms and interactions more adapted to particular living organisms, see for instance [19], [20], [21], [22] and [23].

In this paper, we focus on the cases with velocity-independent interactions. More precisely, by introducing a pairwise symmetric interaction potential $W(x)=U(|x|)$, we consider the two-dimensional model

$$\frac{dx_i}{dt} = v_i, \tag{1}$$

$$\frac{dv_i}{dt} = \alpha v_i - \beta v_i |v_i|^2 - \sum_{j \neq i} \nabla W(x_i - x_j),$$

where $x_i, v_i \in R^2, i = 1, \ldots, N$ are the positions and velocities of the individual particles and α, β are effective values for self-propulsion and friction forces, see [10],[11] and [12] for more discussions.

For this relatively simple system, a variety of patterns are observed, for instance coherent moving flocks and rotating mills [10]. Other patterns like rings and clumps were discovered in [11] and [12], with the introduction of the concept of *H-stability* from statistical physics to better characterize the patterns in the parameter space. Delta rings, with mass uniformly distributed on a circle, have been further studied thoroughly for power-law like potentials, we refer to [24], [25], [26], [27], [28], [29] and [30] for details. While the stability and bifurcation of the ring solutions can be investigated in a straightforward way because of their explicit particle representations on a circle, there are few studies on the more prevalent compact steady solutions, like flocks or mills. The reason lies in the difficulties to solve some complicated integro-differential equations for most of the popular choices of the potential W. Only for certain particular potentials like the quasi-Morse type proposed in [31], the existence and uniqueness of the coherent moving flocks can be established rigorously [32].

In this paper, we focus on the stability of flock solutions of (1) for general potentials defined below.

Definition 1.

A flock solution of the particle model (1) is a spatial configuration $\hat{x}$ with zero net interaction force on every particle, that translates at a uniform velocity $m_0 \in R^2$ with , $|m_0| = \sqrt{\frac{\alpha}{\beta}}$, hence $(x_i(t), v_i(t)) = (\hat{x}_i - tm_0, m_0)$.

We note, that the spatial configuration $\hat{x}$ is a stationary state to the first-order interacting particle system

$$\frac{dx_i}{dt} = -\sum_{j \neq i} \nabla W(x_i - x_j), \tag{2}$$

which is analyzed for example in [8] and [9]. In this work, we focus on general flock solutions whose spatial configurations tend to a compactly supported particle density in the continuum limit.

One conclusion from our analysis is the somehow unexpected deep relation between the linear stability of the flock spatial configuration as steady state for the first-order swarming model (2) and the nonlinear stability of the family of associated flock solutions for the second-order swarming model (1). This relation was already found in the linear stability analysis around flock solutions in [30]. There the authors showed that the linearization of (2) around the equilibrium state $\hat{x}$ has a positive eigenvalue if and only if the linearization of (1) around the steady flock solution in the co-moving frame has an eigenvalue with positive real part. Moreover, they show that if the equilibrium state $\hat{x}$ is linearly stable for (2), then the associated flock solution is always linearly unstable due to the presence of a generalized eigenvector associated to the zero eigenvalue of the linearization of the flock solution in the co-moving frame due to symmetries.

However, it is in this work where we clarify completely this deep imbrication by showing the nonlinear stability of the flock solutions under mild hypotheses on the linearization of(2) about the spatial configuration $\hat{x}$. Actually, the basic hypotheses, apart from few technical assumptions, are that the steady state $\hat{x}$ of (2) is linearly asymptotically stable except the obvious symmetries: translations and rotations. Therefore, we can apply our theorem to verify the nonlinear stability of flock solutions to more biologically adapted potentials such as the Morse and the Quasi-Morse potentials [11].

The main result of this paper asserts, except few technicalities, that *the family of flock solutions* associated to a linearly asymptotically stable steady state of (2) is*asymptotically stable* for the dynamics of (1). Here, the asymptotic stability of the family of flock solutions means that any small enough perturbation in (x,v)-space at any time t_0 will, under the dynamics of the system (1), relax towards (likely) another flock solution in the family at an exponential rate as $t \to \infty$. Let us finally emphasize that the most rigorous way of stating our main theorem uses advanced concepts of dynamical systems [33]. Our main theorem can be rephrased as follows: the family of flock solutions to (1)associated to a linearly (except symmetries) asymptotically stable steady state of (2)forms a normally hyperbolic invariant manifold for the system (1) with an empty unstable manifold.

This result has important applications, especially in the study of flock patterns of (1)using particle simulations. In general, the desired patterns are observed only after certain transit dynamics, and with carefully prepared initial data. With the spatial configurations from the first order system (2) (if they exist), the computational load is reduced significantly, making possible to study the existence of these patterns in great details.

After some preliminary notations and assumptions, the main result of this paper is stated in Section 2. The formulation of the linearized dynamics when the mean velocity is transformed to zero is given in Section 3, where the non-trivial relation between the Jacobian matrices of the two systems (1) and (2) is clarified. A comparison of this work to the techniques used in [30] also is given. In Section 4, we use the eigenspace structures of these Jacobian matrices to reveal the connection between the nonlinear stability of the flock solutions for (1) and the linear stability of their spatial configuration for (2), leading to the proof of the main theorem. Section 5 is devoted to numerical experiments to check the validity of the assumptions on the linear stability for the steady state of the system (2)for different biologically reasonable potentials.

2. MAIN RESULT

In this section, we formulate the main theorem of this work. We will start with some assumptions on the spatial configuration of flock solutions, which is a stationary state of the first-order swarming model (2). The hypotheses assume this spatial configuration to be stable under the dynamics of (2).

Given a steady spatial particle configuration to (2), i.e, a set of particle positions $(\hat{x}_i)_{i=1}^N$such that

$$\sum_{j \neq i} \nabla W(\hat{x}_i - \hat{x}_j) = 0, \quad \text{for all } i \in \{1, \ldots, N\}, \tag{3}$$

we denote by $\hat{x}$ the $2N$-dimensional spatial configuration vector $(\hat{x}_1, \ldots, \hat{x}_N)$. We can now write the linearization of the system (2) at the steady configuration $\hat{x}$ as

$$\frac{\mathrm{d}h}{\mathrm{d}t} = G(\hat{x})h, \tag{4}$$

where $G(\hat{x})$ denotes the Jacobian matrix associated to (2) and $h \in \mathbb{R}^{2N}$. The $2N \times 2N$-Jacobian matrix is explicitly given by $G(\hat{x}) = (G_{ij})$ with G_{ij} being the 2×2-blocks defined as

$$G_{ij} = \begin{cases} -\sum_{j \neq i} \operatorname{Hess} W(\hat{x}_i - \hat{x}_j) & \text{for } i = j \\ \operatorname{Hess} W(\hat{x}_i - \hat{x}_j) & \text{for } i \neq j, \end{cases} \tag{5}$$

with $\operatorname{Hess} W$ denoting the Hessian matrix of W.

We will make extensive use of the following standard notation: given two matrices $A \in \mathbb{R}^{n \times m}$ and $B \in \mathbb{R}^{p \times q}$, the Kronecker product $A \otimes B$ is defined as the matrix $A \otimes B = (a_{ij}B)_{ij} \in \mathbb{R}^{np \times mq}$. Let us also denote by $0_n, 1_n$ the column vectors of length n with all entries equal to 0 or 1 respectively, and I_n the identity matrix of size n. Let us now specify the precise assumptions on $\hat{x}$, and thus indirectly on the potential W.

Assumption 1.

Given an interaction potential W, we assume the existence of a linearly asymptotically stable (except symmetries) stationary state not lying on a line, i.e., a spatial configuration $\hat{x}$ such that the following assumptions hold:

•(H1) $\hat{x}$ is a stationary state of (2), i.e. $\hat{x}$ satisfies (3).

•(H2) The eigenspace for the zero eigenvalue of $G(\hat{x})$ is spanned by the eigenvectors

$$w_1 = 1_N \otimes \begin{pmatrix}1\\0\end{pmatrix}, \quad w_2 = 1_N \otimes \begin{pmatrix}0\\1\end{pmatrix}, \quad \text{and} \quad w_3 = \left(I_N \otimes \begin{pmatrix}0 & -1\\1 & 0\end{pmatrix}\right)\hat{x} \tag{6}$$

representing invariance of $G(\hat{x})$ with respect to translations and rotations in R^2, and therefore is three dimensional, $\dim(\mathrm{Eig}(G(\hat{x}), 0)) = 3$.

•(H3) All other eigenvalues of $G(\hat{x})$ have negative real parts, i.e. $Re(\lambda_i) < 0 \Leftrightarrow \lambda_i \neq 0$ for all $i \in \sigma(G(\hat{x}))$.

•(H4) Not all points of $\hat{x}$ lie on a straight line in R^2.

Hypotheses (H1)–(H3) imply that $\hat{x}$ is linearly asymptotically stable for (4) except for the obvious symmetries, or equivalently that the Jacobian matrix $G(\hat{x})$ is negative definite except for the null eigenspace in (6). As a consequence, $\hat{x}$ is nonlinearly asymptotically stable for the dynamics of (2) except for translations and rotations. More rigorously speaking and making use of more advanced dynamical systems theory [33], properties (H1)–(H3) imply that the family of stationary states of (2) obtained by rotations and translations from $\hat{x}$ given by

$$RT(\hat{x}) = \{y : y = 1_N \otimes b + (I_N \otimes R(\phi))\hat{x}, \phi \in [0, 2\pi), b \in \mathbb{R}^2\}$$

forms a normally hyperbolic invariant manifold for the system of ODEs in (2) for which the unstable manifold is empty. Here,

$$R(\phi) = \begin{pmatrix}\cos\phi & -\sin\phi\\ \sin\phi & \cos\phi\end{pmatrix}$$

is the 2×2 rotation matrix of angle ϕ. This stability concept will be further discussed and used later on for the second order model (1).

Let us remark, that assumptions (H1)–(H4) are very mild in the sense that they are natural properties of stationary states of (2) found through particle simulations from generic initial configurations, e.g. compactly supported flocks. Particular special cases satisfying hypotheses (H1)–(H4) are the so-called Delta rings: particles arranged equidistantly on a circle. We will discuss in the last section more biologically relevant potentials, such as Morse-like potentials [8], [9], [10], [11], [31] and [32], for which we can numerically find stationary states $\hat{x}$ satisfying hypotheses (H1)–(H4).

Now, let us make precise the concept of the family of flock solutions of (1) associated to the stationary state $\hat{x}$ of (2). Let us denote by V_s the set of possible steady velocity vectors for a flock solution of speed $s = \sqrt{\frac{\alpha}{\beta}}$ to (1) defined as $V_s=\{m_0 \in R^2 : |m_0| = s\}$. Then the family of flock solutions of (1) associated to the steady state $\hat{x}$ of (2) is given by

$$Z_F = \left\{ z = \begin{pmatrix} x^* + v^* t \\ v^* \end{pmatrix}_{t \in \mathbb{R}}, v^* = 1_N \otimes m_0, x^* \in RT(\hat{x}), m_0 \in \mathcal{V}_s \right\}. \quad (7)$$

It is straightforward to check that all $z \in Z_F$ are solutions to (1) in vectorial form. The next section will be devoted to analyse the system (1) in suitable coordinates to study their stability. A change of variables to (1) that eliminates the translation of one particular member of the flock family is given by the co-moving frame $\tilde{v}_i := v_i - m_0$ transforming the flock (x^*+v^*t, v^*) onto a stationary state. However, it is very natural to think of small perturbations on (x^*+tv^*, v^*) that, under the dynamics of (1), tend to possibly another spatial configuration in $RT(\hat{x})$ translating at a different velocity generated by $\tilde{m}_0$, i.e. to another member of the family of flocks Z_F. Hence, we cannot expect linear stability for general perturbations for such a change of variables.... Instead, we propose a new change of variables measuring the difference to mean velocity in velocity space, and aim to establish nonlinear stability in the *entire family* Z_F.

Definition 2.

Defining the mean velocity of N particles as

$$m(t) := \frac{1}{N}\sum_{i=1}^{N} v_i(t),$$

we introduce the change of variables with respect to the mean velocity $m(t)$ as:

$$\tilde{x}_i(t) := x_i(t) - \int_0^t m(s)ds, \tag{8}$$

$$\tilde{v}_i(t) := v_i(t) - m(t).$$

We will work with the dynamical system (1) and its linearization in vector form, and thus we need to introduce some matrix notation used in the following.

Definition 3 Notation.

1.$0_{n\times p}$ is a block of zeros of size $n \times p$.

2.Writing $\delta_n^i \in \mathbb{R}^n$, we refer to a vector of length n with value 1 as its i-th entry and zeros elsewhere.

3.For the sake of simplicity, we occasionally write column vectors as row vectors neglecting the transpose operation on its components, e.g.

$$(x, v, 0_2)^T := (x^T, v^T, 0_2^T)^T = \begin{pmatrix} x \\ v \\ 0_2 \end{pmatrix}.$$

4.For a square matrix $A \in R^{n\times p}$ or a column vector $b \in R^n$, we denote by

$$\lceil A \rceil := (a_{ij})_{\substack{i=1,\dots,n-2,\\ j=1,\dots,p}} \qquad \lceil b \rceil = (b_i)_{i=1,\dots,n-2},$$

the matrix obtained from A by eliminating the last two rows and analogously for vectors. The k-th row or column of A is labelled as $a_{k,\cdot}$ or $a_{\cdot,k}$ respectively.

Lemma 1.

Using the change of coordinates (8), *the microscopic model* (1) *reads*

$$\frac{d\tilde{x}_i}{dt} = \tilde{v}_i,$$

$$\frac{d\tilde{v}_i}{dt} = \frac{N-1}{N}\left(\alpha - \beta|\tilde{v}_i + m|^2\right)(\tilde{v}_i + m) - \frac{1}{N}\sum_{j\neq i}\left(\alpha - \beta|\tilde{v}_j + m|^2\right)(\tilde{v}_j + m) - \sum_{j\neq i}\nabla W(\tilde{x}_j - \tilde{x}_i), \tag{9}$$

$$\frac{dm}{dt} = \frac{1}{N}\sum_j\left(\alpha - \beta|\tilde{v}_j + m|^2\right)(\tilde{v}_j + m).$$

For any mean velocity $m_0 \in V_s$, *the vector* $\hat{Q} = (\hat{x}, 0_{2N}, m_0)^T$ *is a stationary state of* (9). *Moreover, the Jacobian matrix* F *obtained by linearizing* (9) *around a stationary state* $\hat{Q}$ *with an arbitrary orientation* m_0 *is given by*

$$F = \begin{pmatrix} 0_{2N\times 2N} & I_{2N} & 0_{2N\times 2} \\ G(\hat{x}) & -\begin{pmatrix} \frac{N-1}{N} & -\frac{1}{N} & \cdots & -\frac{1}{N} \\ -\frac{1}{N} & \ddots & & \vdots \\ \vdots & & \ddots & -\frac{1}{N} \\ -\frac{1}{N} & \cdots & -\frac{1}{N} & \frac{N-1}{N} \end{pmatrix} \otimes 2\beta(m_0 \otimes m_0^T) & 0_{2N\times 2} \\ 0_{2\times 2N} & -1_N^T \otimes \frac{2\beta}{N}(m_0 \otimes m_0^T) & -2\beta(m_0 \otimes m_0^T) \end{pmatrix}. \tag{10}$$

Proof.

By definition $\frac{d\tilde{x}_i}{dt} = \frac{dx_i}{dt} - m(t) = v_i(t) - m(t) = \tilde{v}_i(t)$. Next we note $\tilde{x}_i - \tilde{x}_j = x_i - x_j$, and hence $\nabla W(\tilde{x}_i - \tilde{x}_j) = \nabla W(x_i - x_j)$. Now, we compute

$$\begin{aligned}\frac{dm}{dt} &= \frac{1}{N}\sum_i \frac{dv_i}{dt} = \frac{1}{N}\left(\sum_i (\alpha - \beta|v_i|^2)\, v_i + \sum_i \sum_{j\neq i} \nabla W(x_i - x_j)\right) \\ &= \frac{1}{N}\sum_i \left(\alpha - \beta|\tilde{v}_i + m|^2\right)(\tilde{v}_i + m),\end{aligned}$$

using that the net total force in the system is zero by symmetry, i.e. Newton's action–reaction principle. From this, the equation for $\frac{d\tilde{v}_i}{dt} = \frac{dv_i}{dt} - \frac{dm}{dt}$ is readily available. Now, any state $\hat{Q} = (\hat{x}, 0_{2N}, m_0)^T$, with $m_0 \in V_s$ is a stationary state of (9) since $G(\hat{x}) = 0$ and all self-propulsion terms vanish. Let us drop the tildes henceforth for simplicity and let us compute the Jacobian matrix of the system (9). The first line of blocks in F is trivial. In the second line of blocks, we first have $G(\hat{x})$ since the dependency on x in (9)is identical to (1), hence one obtains the Jacobian of the first-order particle system (2)evaluated at the stationary spatial configuration. To compute the second block, consider an arbitrary particle i and its velocity vector $(v_{i,1}, v_{i,2})^T$. Then, identifying the right-hand sides of (9) with the time derivative symbols, we get:

$$\begin{aligned}\partial_{v_{i,1}} \frac{dv_{i,1}}{dt} &= \frac{N-1}{N}\left(-2\beta(v_{i,1} + m_1)^2 + (\alpha - \beta|v_i + m|^2)\right), \\ \partial_{v_{i,2}} \frac{dv_{i,1}}{dt} &= \frac{N-1}{N}\left(-2\beta(v_{i,2} + m_2)(v_{i,1} + m_1) + (\alpha - \beta|v_i + m|^2)\right), \\ \partial_{v_{j,1}} \frac{dv_{i,1}}{dt} &= -\frac{1}{N}\left(-2\beta(v_{j,1} + m_1)^2 + (\alpha - \beta|v_j + m|^2)\right),\end{aligned}$$

for $j \neq i$ and analogous expressions for $\partial_{v_{i,1}} \frac{dv_{i,2}}{dt}$, $\partial_{v_{i,2}} \frac{dv_{i,2}}{dt}$, $\partial_{v_{j,2}} \frac{dv_{i,1}}{dt}$, $\partial_{v_{j,1}} \frac{dv_{i,2}}{dt}$, and $\partial_{v_{j,2}} \frac{dv_{i,2}}{dt}$. Evaluating at $\hat{Q}$ the last expression in all terms vanishes since $|m_0|^2 = \frac{\alpha}{\beta}$, which in total gives the structure of the block using the Kronecker product notation. In the third line ofF, the second block follows analogously. For the last block, we compute

$$\partial_{m_1}\frac{dm_1}{dt}=\frac{1}{N}\sum_j\left(-2\beta(v_{j,1}+m_1)^2+(\alpha-\beta|v_j+m|^2)\right),$$

which reduces to $\partial_{m_1}\frac{dm_1}{dt}=-2\beta m_1^2$ at $\hat{Q}$, and analogously for the remaining partial derivatives. This completes the computation of the structure of F.

Let us point out that the family of flock solutions to (1) is translated through the change of variables (8) to the set of stationary solutions of (9) given by

$$\tilde{Z}_F=\left\{Q^*=\begin{pmatrix}x^*\\0_{2N}\\m_0\end{pmatrix},x^*\in RT(\hat{x}),m_0\in\mathcal{V}_s\right\}. \tag{11}$$

Now, we have almost all the ingredients to write the main result of this paper that holds under an additional technical assumption on the interplay between the linearization and the orientation of the flock.

Assumption 2 H5.

All eigenvectors to non-zero eigenvalues of $G(x^*)$ are not particle-wise orthogonal to the fixed mean-velocity m_0. This means, that for pairs of entries $2i-1, 2i$ of any eigenvector w of $G(x^*)$, we have

$$\left\langle m_0, w_{\substack{2i-1\\2i}}\right\rangle\neq 0\quad\text{for at least one } i\in 1,\ldots,N.$$

We will understand the role of this additional assumption below, but let us point out that this assumption is generically met by the members of the flock family $\tilde{Z}_F$. The main result of this work reads as:

Theorem 1.

Assume that the symmetric pairwise interaction potential W *allows for a stationary spatial configuration* $\hat{x}$ *of* (2) *satisfying assumptions* (H1)–

(H4) . *Consider the second-order swarming model* (1) *with* $\alpha,\beta>0$, *then the family of flock solutions* Z_F *of* (1) *defined in* (7) *is locally asymptotically stable for all configurations* x^* *satisfying* (H5) *in the following sense: Any small enough perturbation in* (x,v)*-space at any time* t_0 *of the flock solution* $z^* \in Z_F$ *associated to* x^* *will, under the dynamics of the system* (1), *relax towards another flock solution* $\tilde{z} \in Z_F$ *at an exponential rate as* $t \to \infty$.

Let us remind that assumptions (H1)–(H4) imply in particular that the family of stationary states $RT(\hat{x})$ forms a normally hyperbolic invariant manifold for (2) with empty unstable manifold. Therefore, our main theorem can be rephrased as stating that the family of flock solutions associated to $\hat{x}$ of the second order model (1) has essentially the same property. More precisely, we have the following result.

Corollary 1.

Assume that the symmetric pairwise interaction potential W *allows for a stationary spatial configuration* $\hat{x}$ *of* (2) *satisfying assumptions* (H1)–(H4) . *Then the subset of stationary states of* $\tilde{Z}_F$ *in* (11) *satisfying* (H5) *is a normally hyperbolic invariant manifold of* (9) *with empty unstable manifold.*

Next two sections are devoted to the proof of these results. We start in next section by restricting the dynamics to a physically consistent subspace.

3. PHYSICALLY CONSISTENT DYNAMICS AND ITS LINEARIZATION

Note that (9) is a $4N+2$ dimensional system and thus $F \in R^{(4N+2)\times(4N+2)}$ by having added the mean velocity as an additional variable. However, a stability analysis of (10)would include unphysical perturbations where particle velocities do not match mean velocity. The dynamics we are interested in are $4N$-dimensional, and thus we need to reduce (10) to physically consistent dynamics. Clearly, mean velocity consistent states are invariant under the dynamics of (9), since

$$\frac{\mathrm{d}}{\mathrm{d}t}\frac{1}{N}\sum_i (v_i - m) = 0.$$

We hence define a base for the 4N-dimensional subspace of mean velocity consistent states by removing the redundant equation of the N-th particle.

Definition 4.

Define the subspace of mean velocity consistent states as span(B) with $B=\{b_i\}_{i=1,\dots,4N}$ and

$$b_i = \begin{cases} (\delta^i_{2N}, & 0_{2N}, & 0_2)^T & i = 1,\dots,2N, \\ (0_{2N}, & \delta^k_{2N-2} \; -1 \; 0, & 0_2)^T & i = 2N + (2k-1),\, k = 1,\dots,N-1, \\ (0_{2N}, & \delta^k_{2N-2} \; 0 \; -1, & 0_2)^T & i = 2N + 2k,\, k = 1,\dots,N-1, \\ (0_{2N}, & 0_{2N}, & \delta^k_2)^T & i = 4N - 2 + k,\, k = 1,2. \end{cases}$$

Any physically relevant perturbation of (1) is projected onto span(B) via

$$P : \begin{pmatrix} \Delta x \\ \Delta v \end{pmatrix} \mapsto \begin{pmatrix} \Delta x \\ \hline \Delta v_1 - \frac{1}{N}\sum \Delta v_i \\ \vdots \\ \Delta v_N - \frac{1}{N}\sum \Delta v_i \\ \hline \frac{1}{N}\sum \Delta v_i \end{pmatrix} \in \mathrm{span}(B).$$

Clearly, B is isomorphic to a base of R^{4N} and the relevant invariant subspace for the dynamics of (9). As mean velocity consistency is a linear algebraic constraint to (9), it turns out that span(B) is also an invariant subspace of F, which means we can restrict the stability analysis of (10) to physically relevant perturbations by computing the matrix representation of F:span(B)→span(B) under a change of base to B, denoted by F^B_B.

Lemma 2.

span(B) *is an invariant subspace of* F. *Restricting* F *to* span(B) *and changing base to* B *gives*

$$F_B^B = \begin{pmatrix} 0_{2N\times 2N} & \begin{matrix} I_{2N-2} \\ -1_{N-1}^T \otimes I_2 \end{matrix} & 0_{2N\times 2} \\ \lceil G(\hat{x}) \rceil & -I_{N-1} \otimes 2\beta(m_0 \otimes m_0^T) & 0_{2N-2\times 2} \\ 0_{2\times 2N} & 0_{2\times 2N-2} & -2\beta(m_0 \otimes m_0^T) \end{pmatrix}, \quad (12)$$

where $\lceil \cdot \rceil$ *is used according to Definition* 3.4.

Proof.

We compute the image of base vectors b_i under F. First, consider b_i with a non-zero entry in a spatial coordinate $(i=1,\dots,2N)$. Then $Fb_i=(0_{2N}, g_{\cdot,i}, 0_2)^T$ consists of the i-th column of $G(\hat{x}) = (g_{i,j})_{ij}$. Since the sum of each column of $G(\hat{x})$ in (5) is zero, we can rewrite this in terms of B as

$$(0_{2N}, g_{\cdot,i}, 0_2)^T = \sum_{l=1}^{2N-2} g_{l,i} b_{2N+l},$$

which gives the first block column of (12). Next, consider b_i with a non-zero entry in a velocity component k and assume i to be odd $(i=2N+(2k-1), k=1,\dots,N-1)$. Then we can compute Fb_i to get

$$\begin{pmatrix} \begin{matrix} \delta_{2N-2}^k \\ -1 \\ 0 \end{matrix} \\ \hline -\begin{pmatrix} \frac{N-1}{N} & \cdots & -\frac{1}{N} \\ \vdots & \ddots & \vdots \\ -\frac{1}{N} & \cdots & \frac{N-1}{N} \end{pmatrix} \otimes 2\beta(m_0 \otimes m_0^T) \begin{pmatrix} \delta_{2N-2}^k \\ -1 \\ 0 \end{pmatrix} \\ \hline 0_2 \end{pmatrix} = \begin{pmatrix} \delta_{2N-2}^k \\ -1 \\ 0 \\ \hline 0_{2(k-1)} \\ -2\beta(m_0 \otimes m_0^T)_{\cdot,1} \\ 0_{2(N-k)-2} \\ 2\beta(m_0 \otimes m_0^T)_{\cdot,1} \\ \hline 0_2 \end{pmatrix}$$

since, in the second block, the multiplication with $(\delta^k_{2N-2}, -1, 0)^T$ first gives $\frac{N-1}{N} + \frac{1}{N}$ in row k, $-\frac{1}{N} - \frac{N-1}{N}$ in row 2 N − 1 and zero in all other rows, which is then multiplied with $-2\beta(m_0 \otimes m_0^T)$ to select its first column twice (as i is odd), indicated by the subscript$()_{\cdot,1}$. This rewrites in terms of B as

$$Fb_i = b_{i-2N} - b_{2N-1} - 2\beta m_{0,1}^2 b_i - 2\beta m_{0,1} m_{0,2} b_{i+1}.$$

Analogously, one obtains for even i (i = 2 N + 2 k, k = 1, ..., N − 1)

$$Fb_i = b_{i-2N} - b_{2N} - 2\beta m_{0,1} m_{0,2} b_{i-1} - 2\beta m_{0,2}^2 b_i,$$

involving the second column of $-2\beta(m_0 \otimes m_0^T)$. The change of base for vectors of B with non-zero entry in the mean velocity component is trivial. Having shown that all images of base vectors can be expressed in terms of B, we showed the invariant subspace property and obtain the block structure of F_B^B from our computations.

We now derive two corollaries on the linear structure of F_B^B, which will be essential to establish our stability result. To do so, we first rewrite (12) as

$$F_B^B =: \begin{pmatrix} H & 0_{4N-2\times 2} \\ 0_{2\times 4N-2} & -2\beta(m_0 \otimes m_0^T) \end{pmatrix}, \tag{13}$$

since the two blocks separate. We first investigate the eigenspace associated to the zero eigenvalue of F_B^B. ...

Lemma 3.

Suppose that $G(\hat{x})$ *satisfies* (H1)–(H4) , *then* F_B^B *does not possess a generalized eigenvector for eigenvalue zero.*

Proof.

Since the lower 2×2 block of F_B^B in (13) can be diagonalized with eigenvalues 0 and -2α, then we restrict our analysis to H. To simplify the notation, we will skip the dependency of the matrix $G(\hat{x})$ on the stationary state in this proof. If $z=(x,v)^T \in Eig(H,0)$ is a zero eigenvector, then $v=0_{2N-2}$ and $x \in Eig(G,0)$ by (12). Suppose $(u,w)^T$ is a generalized eigenvector, then

$$H\begin{pmatrix} u \\ w \end{pmatrix} = \begin{pmatrix} x \\ 0_{2N-2} \end{pmatrix}, \quad x \in \text{Eig}(G, 0)m,$$

which is equivalent to

$$w = \lceil x \rceil, \tag{14a}$$

$$\sum_{i=1}^{N} x_{2i-1} = \sum_{i=1}^{N} x_{2i} = 0, \tag{14b}$$

$$\lceil Gu \rceil - I_{N-1} \otimes 2\beta(m_0 \otimes m_0^T)w = 0_{2N-2}. \tag{14c}$$

Due to the definition of G in (5), we get

$$\sum_{i=1}^{N} g_{2i-1,\cdot} = \sum_{i=1}^{N} g_{2i,\cdot} = 0_{2N}^T. \tag{15}$$

Adding two rows using (15), (14a), and (14b), (14c) is equivalent to

$$Gu - I_N \otimes 2\beta(m_0 \otimes m_0^T)x = 0_{2N}. \tag{16}$$

We left-multiply with x and get

$$0 = x^T G u = x^T (I_N \otimes 2\beta(m_0 \otimes m_0^T))x, \qquad (17)$$

since G symmetric and $x \in \mathrm{Eig}(G,0)$. This implies $x = [a_1 \dots a_N]^T \otimes m^\perp$ for some constants a_i since the right-sided expression is quadratic and hence x has to be particle-wise orthogonal to m_0 for the equation to hold ($m_0^\perp := (-m_{0,2}, m_{0,1})^T$). On the other hand, (14b) implies that $x \in \mathrm{span}(w_3)$, since w_3, as given in (6), is the only zero eigenvector of G whose component sums can be zero. Hence, all position vectors are tangential to a fixed mean velocity m_0, which is a contradiction to (H4).

... Next, we investigate the eigenvalues of F_B^B under another hypothesis linking the eigenvectors of $G(\hat{x})$ with the orientationally arbitrary but fixed mean velocity m_0.

Lemma 4.

Suppose $G(\hat{x})$ *satisfies* (H1)–(H4) *and together with* m_0 *satisfies* (H5) *, then* $\dim(\mathrm{Eig}(F_B^B, 0)) = 4$*. Furthermore, all non-zero eigenvalues* μ_i *of* F_B^B *have negative real-part,* $\mu_i \neq 0 \Rightarrow \mathrm{Re}(\mu_i) < 0$.

Proof.

Lemma 3 shows that there are no generalized eigenvectors for the zero eigenvalue of F_B^B. We will again skip the dependency of the matrix G on the stationary state. Let us first show that $\dim(\mathrm{Eig}(F_B^B, 0)) \geq 4$. One directly checks that the basis elements $\{w_1, w_2, w_3\}$ of $\mathrm{Eig}(G,0)$ define linearly independent eigenvectors in $\mathrm{Eig}(F_B^B, 0)$ by $z_i := (w_i, 0_{2N-2}, 0_2)^T, i = 1,2,3$. As remarked in Lemma 3, the lower 2×2 block of F_B^B can be diagonalized with eigenvalues 0 and -2α and associated eigenvectors $z_4 := (0_{2N}, 0_{2N-2}, m_0^\perp)^T$ and $(0_{2N}, 0_{2N-2}, m_0)^T$ respectively.

Next, we prove the last statement and $\dim(\mathrm{Eig}(F_B^B, 0)) \leq 4$ simultaneously. First, note that any pair of eigenvalue μ and eigenvector $z = (x, v)^T$ satisfies $Hz = \mu z$, which gives

$$v = \mu \lceil x \rceil \tag{18a}$$

$$\sum_{i=1}^{N} x_{2i-1} = \sum_{i=1}^{N} x_{2i} = 0 \tag{18b}$$

$$\lceil Gx \rceil - (I_{N-1} \otimes 2\beta (m_0 \otimes m_0^T))v = \mu v. \tag{18c}$$

Eq. (18c) rewrites as

$$\mu^2 \lceil x \rceil + (I_{N-1} \otimes 2\beta \mu (m_0 \otimes m_0^T)) \lceil x \rceil = \lceil Gx \rceil. \tag{19}$$

By the same argument used in Lemma 3, (19) is equivalent to

$$\mu^2 x + (I_N \otimes 2\beta \mu (m_0 \otimes m_0^T))x = Gx, \tag{20}$$

due to (18b) and the properties of G. We left-multiply with $\bar{x}$ to obtain

$$\mu^2 + 2\beta\mu \sum_{i=1}^{N} \left\langle m_0, x_{\substack{2i-1\\2i}} \right\rangle \overline{\left\langle m_0, x_{\substack{2i-1\\2i}} \right\rangle} = \bar{x}^T Gx, \tag{21}$$

assuming without loss of generality the normalization $\langle x, x \rangle = \bar{x}^T x = 1$ and denoting the complex scalar product as $\langle \cdot,\cdot \rangle$. Therefore, we deduce

$$\mu = -A \pm \sqrt{A^2 - B} \quad \text{with } A = -\beta \sum_{i=1}^{N} \left| \left\langle m_0, x_{\substack{2i-1\\2i}} \right\rangle \right|^2 \text{ and } B = -\bar{x}^T Gx.$$

Note that $A \geq 0, B \geq 0$ since $\bar{x}^T Gx \leq 0$ for any x due to (H3). We study three cases:

Case 1: Suppose $A > 0, B > 0$. If $A^2 - B \geq 0 \Rightarrow \mu < 0$ by monotonicity. If $A^2 - B < 0$, its square root is purely imaginary and thus $Re(\mu_i) < 0$.

Case 2: Suppose $A = 0, B > 0$. Then $\mu = \pm\sqrt{\bar{x}^T Gx} \in i\mathbb{R}$. Since A is a sum over squared moduli of N complex numbers, $A = 0 \Leftrightarrow \left\langle m_0, x_{\frac{2i-1}{2i}} \right\rangle = 0$ for all i. Since m is fixed, this holds if and only if x is particle-wise orthogonal to m_0. Now, inserting $A = 0$ back into (20)gives $Gx = \mu^2 x, \mu^2 < 0$ which implies that such an x is also an eigenvector for G. This is a contradiction to (H5) and no such case exists.

Case 3: Suppose $A \geq 0, B = 0$. Then $x \in Eig(G, 0)$ and either $\mu = -2A < 0$ or $\mu = 0$. For the second case, $v = 0_{2N-2}$ follows from (18a), which implies $z \in \text{span}(z_1, z_2, z_3) \subset \text{Eig}(F_B^B, 0)$ by (H2).

Together, this shows that

$$\mu_i \neq 0 \Rightarrow \text{Re}(\mu_i) < 0. \quad \text{(22a)}$$

$$A \geq 0, B = 0, \mu = 0 \Rightarrow z \in \text{span}(z_1, z_2, z_3) \subset \text{Eig}(F_B^B, 0). \quad \text{(22b)}$$

To complete the study of $\text{Eig}(F_B^B, 0)$ suppose now that $\mu = 0$. Then $A^2 - B \geq 0$ and $0 = -A + \sqrt{A^2 - B} \Leftrightarrow A^2 = A^2 - B \Leftrightarrow B = 0$, which takes us back (22b). Therefore, no further eigenvector to eigenvalue zero exists, which completes the proof.

Let us remark that the 4 basis eigenvectors associated to the zero eigenvalue of F_B^B represent linearized flow within the set of stationary flock solutions $\tilde{Z}_F$ in (11). From the change of variables (8) and Lemma 1, we know that the family of flock solutions is written as $\hat{Q} = (\hat{x}, 0_{2N-2}, m_0)^T$ with $|m_0|^2 = \frac{\alpha}{\beta}$. $G(\hat{x})$ is invariant under translation and spatial rotation, and thus z_1, z_2, and z_3 are the linearized flow along these invariances. The fourth eigenvector z_4 represents linearized rotation of the mean velocity m_0....

Remark 1 Comparison to the Techniques of [30].

The decisive technical difference used in the analysis presented here compared to [30]is the change of variables with respect to mean velocity. Here, we are able to show the absence of a generalized eigenvector in Lemma 3 and Lemma 4. In the co-moving frame used in [30], this is not possible as a particular mean velocity is fixed, therefore the instability of a flock solution in the co-moving frame is shown. Instead, the instability criterion are in fact established proving the existence of a generalized eigenvectors associated to the zero eigenvalue (see [30, Remark 3.6]). The difference of the two approaches can also be seen in the proof of Lemma 4, which bears some technical similarity with [30, Lemma 3.1], but additionally uses the mean-velocity consistency in(18b) and (19), (20).

4. PROOF OF THE MAIN RESULT

Changing coordinates to (8) eliminates the linear translation at constant speed and flock solutions become constant solutions which we identify with their profile and velocity at $t=0$. The image of Z_F under (8) is $\tilde{\mathcal{Z}}_F$ in (11). $\tilde{\mathcal{Z}}_F$ is a manifold of stationary states of the flow of system (9). We aim to show that $\tilde{\mathcal{Z}}_F$ is a normally hyperbolic invariant manifold according to [33] for the time flow map of (9) restricted on $\mathrm{span}(B)$, which we denote by $\mathbb{F}^t_B$. First, $\tilde{\mathcal{Z}}_F$ is trivially invariant, $\mathbb{F}^t_B(\tilde{\mathcal{Z}}_F) = \tilde{\mathcal{Z}}_F$ for all $t \in R$, as it is entirely stationary byLemma 1. Choose $Q^* \in \tilde{\mathcal{Z}}_F$ such that (H5) is fulfilled. Since eigenvectors are distinct, (H5) holds true for a neighbourhood of Q^*. There is no tangential flow along $\tilde{\mathcal{Z}}_F$, and in particular

$$\frac{\mathrm{d}}{\mathrm{d}t} D_Q \mathbb{F}^t_B(Q^*) = F^B_B \left[D_Q \mathbb{F}^t_B(Q^*) \right]. \tag{23}$$

Lemma 3 and Lemma 4 tell us, that F^B_B decomposes into the direct sum of $\mathrm{Eig}(F^B_B, 0)$, which equals the tangential space of $\tilde{\mathcal{Z}}_F$ at Q^*, and a subspace of eigenvectors whose eigenvalues have negative real part and span the non-tangential subspace of $\tilde{\mathcal{Z}}_F$ at Q^*. It is important to stress, that this splitting is preserved under the flow $\mathbb{F}^t_B$, since Q^* is stationary. Adapting from [33], all prerequisites for $\tilde{\mathcal{Z}}_F$ to be a normally hyperbolic invariant manifold are given. In the language of dynamical systems theory, the tangent bundle of R^{4N} at Q^* has an F_B-invariant splitting into tangential and stable sub-bundles. A necessary inequality between the tangential flow and the stable decay rate is trivially given in our case, since flow along $\tilde{Z}_F$ is zero and as a consequence, we have a spectral gap. Note, that though $\tilde{Z}_F$ is not compact we can easily select a compact sub-manifold around Q^* or perform a symmetry reduction and eliminate the translational invariance by further reducing $span(B)$.

We can hence apply the theorem of Hadamard and Perron as in [33, Theorem 4.1], which, adapted to our situation, gives the following: There exists locally an F_B-invariant stable smooth manifold in a neighbourhood of $\tilde{Z}_F$, where the splitting into stable and stationary of the linearized flow is preserved. The space between both manifolds is laminated by invariant fibres tangent to the stable subspace at Q^*, which are curves of the nonlinear flow towards $\tilde{Z}_F$. Hence, for a small perturbation of Q^* within that laminated space, we are sure to be sitting on some fibre, which might have Q^* as its base point on $\tilde{Z}_F$ or a point in a neighbourhood of Q^*. In either case, there is transport along the fibre towards $\tilde{Z}_F$, for which an exponential decay estimate holds. This estimate is obtained by closing (23) on the stable sub-bundle. In other words, there exists locally a decomposition into fast and slow (stationary, in our case) coordinates of the nonlinear flow, such that the dynamics is characterized by the linearized flow. Going back to original coordinates (1), we obtain the statements in Theorem 1 and Corollary 1.

Remark 2.

The precise meaning of "relaxation in time" towards a different flock solution in the statement of Theorem 1 is given by the notion of stable hyperbolicity of the family of flock solutions in coordinates (8), as used in the proof. In the original coordinates, there is no notion of "closeness" between elements of (7), since e.g. a change of orientation will cause two flocks to arbitrarily diverge in euclidean norm as $t \to \pm\infty$. Relaxation thus means, that the trajectory of a perturbed flock will approach a trajectory of another flock solution with typically different orientation and similar spatial configuration. Theorem 1also tells us that the manifold is stable towards small perturbations of the flow F_B, though that case is not considered here. Note, that this allows for flock stability under small perturbations of the potential.

5. NUMERICAL EXPERIMENTS

In this section, we numerically investigate the hypothesis of Theorem 1 for a number of commonly used interaction potentials and study the nonlinear stability of flocks in terms of their polarization.

The key advantage of Theorem 1 is that essentially linear (asymptotic except symmetries) stability of the spatial configuration of a flock in the first-order aggregation model (2) is inherited nonlinearly by the family of flock solutions to the second-order model (1). In order to demonstrate that the stability theorem applies to a given potential, (H1)–(H4) have to hold for $\hat{x}$ being a stationary state of (2). We compute a numerical approximation $\{\hat{x}\}$ of $\hat{x}$ by a standard fourth-order Runge–Kutta time integration of (2)with random initial data and an interaction potential scaled with a factor D to relate the amplitude of the right-hand side of (2) with the accuracy of the solver:

$$W_D(r) := D\,W(r).$$

Note, that $\hat{x}$ is independent of D. We hence obtain a spatial configuration for which (H1) holds numerically. Let us mention that (H4) is clearly satisfied for all potentials considered here. Hypotheses (H2)–(H3) require the Jacobian $G(\{\hat{x}\})$ to possess negative eigenvalues except for three zero eigenvalues associated to translation and rotation. As $G(\{\hat{x}\}) \in \mathbb{R}^{2N\times 2N}$ is symmetric it is sufficient to apply standard methods to compute its spectrum, at least for the number of particles we consider. Let$(\mu_1,\mu_2;\mu_3)$ denote the eigenvalues of $G(\{\hat{x}\})$ associated to translation and rotation and $\mu_i, i\geq 4$ all other eigenvalues sorted in descending order. In Table 1 we show μ_4 for different potentials and numbers of particles, together with $|\mu_3|$ to illustrate the numerical accuracy of $\{\hat{x}\}$ and the scaling parameter D. For the comparability of the potentials, D is set such that $\mu_{2N}=1$ for all cases shown. The potentials used inTable 1 are:

	$W(r) = V(r) - CV(r/\ell)$
Morse	$V(r) = -e^{-r}$
Quasi-Morse	$V(r) = -\frac{1}{2\pi}K_0(kr)$
Generalized Morse	$V(r) = -e^{-\frac{r^p}{p}}$
Log-Newtonian	$W(r) = r^2 - \log(r)$

where K_0 is the modified Bessel function of second kind (for detailed discussions of the respective potentials see [32], [31], [12] and [34]).

Table 1. Numerical results for the largest non-zero eigenvalue μ_4 of the Jacobian $G(\{\hat{x}\})$ for varying potentials W and number of particles N. The numerical value of the zero eigenvalue μ_3 is shown as a measure of the exactness of $\{\hat{x}\}$. The free scaling parameter D is set to normalize the smallest eigenvalue to $\min(\mu_i)=-1$. As μ_4 is negative, hypotheses (H2)–(H3) of Theorem 1 are numerically validated for the cases shown.

	N	Eigenvalue structure of G		
		μ_4	$\lvert\mu_3\rvert$	D
Morse potential $C=\frac{10}{9}, \ell=\frac{3}{4}$	25	−6.9579e−5	5.3279e−9	9.5290
	40	−3.9449e−4	3.0435e−8	9.7423
	70	−2.3600e−4	6.8493e−9	10.0008
	100	−2.7069e−5	2.7073e−8	10.0720
Quasi-Morse potential $C=\frac{10}{9}, \ell=\frac{3}{4}, k=\frac{1}{2}$	25	−6.7374e−4	1.4273e−11	39.8647
	40	−1.6636e−4	3.5242e−12	39.0230
	70	−0.0024	9.1205e−12	38.5022
	100	−0.0023	4.0528e−11	38.3616
Generalized Morse potential $C=\frac{10}{9}, \ell=\frac{3}{4}, p=1.25$	25	−5.0193e−5	4.5907e−8	7.6429
	40	−2.7459e−4	1.8389e−8	7.8653
	70	−1.2202e−4	1.7075e−10	7.9694
	100	−8.5882e−6	1.4491e−10	7.9813
Log-Newtonian potential	25	−3.6270e−4	4.0557e−6	0.25
	40	−1.8307e−8	8.0974e−9	0.25
	70	−0.0022	6.1172e−7	0.25
	100	−0.0029	7.1695e−6	0.25

It can be seen that for most cases considered, eigenvalues other than μ_1, μ_2, μ_3 are negative and hence Theorem 1 applies. The difference between the numerical eigenvalue μ_3, which is equal to zero in theory, and μ_4 is of order four or higher, which shows that the negativity of μ_4 is not a numerical artefact. However, in some rare particular cases this conclusion cannot be drawn as μ_3 and μ_4 are of the same order (see the case $N=40$, Log-Newtonian). We remark, that the eigenvalue computations ofTable 1 require a highly accurate computation of $\{\hat{x}\}$, which is here obtained with a scaling of $D=500$ and a time interval $T=[0,500]$.

Next, we aim to study the stability of a flock in the nonlinear dynamics of (1) for the Morse potential under uniformly distributed perturbations in space and velocity. Theorem 1provides a theoretical bound on the decay rate into equilibrium, which is very small as seen in Table 1. For random perturbations it is natural to assume that the flock is "more stable" than the theorem asserts. To study whether a flock remains a flock after an initial perturbation and the subsequent dynamics of (1), we use the notion of polarization of a group of individuals, which is defined as

$$\mathrm{Pol}(\{(x_i, v_i)\}_{i=1}^N) := \frac{1}{N}\sum_i \frac{v_i}{\|v_i\|},$$

for N agents with phase-space coordinates at $(x_i, v_i) \in \mathbb{R}^4$, see e.g. [2] for instance. For an aligned flock, the polarization equals one whereas e.g. a rotating mill possesses zero polarization. We perform a sequence of simulations of (1) with randomly perturbed initial data. The stationary flock solution $z = \{(\hat{x}_i, m_0)\}_{i=1}^N, \|m_0\| = \frac{\alpha}{\beta}$ is perturbed by $\{(\Delta x_i, \Delta v_i)\}_{i=1}^N$ with $(\Delta x_i, \Delta v_i)$ sampled from the uniform distribution on $[-\frac{a}{2}, \frac{a}{2}] \times [-\frac{a}{2}, \frac{a}{2}]$ and varying perturbation strength a. Stability of the flock in the time evolution $[0,T]$ can estimated as follows: If the support of the solution remains bounded and the polarization is equal or larger than the change in polarization induced by the initial perturbation, is it reasonable to assume that the ensemble will again tend to form a flock solution. If however, the polarization drops to significantly low or near zero values, the particles have evolved into an unpredictable state whose convergence behaviour as $T \to \infty$ is unknown. In Fig. 1(a), we plot the initial polarization induced by the perturbation against the minimal perturbation that occurred during a time integration of (1) with $T = 100$. The figure is based on 25,000 simulations with uniformly distributed perturbations strength a and shows the statistical mean and the 5% tails of the statistical distribution. In Fig. 1(b), we show the correlation between the polarization induced by the initial perturbation and the averaged l_2-norm of the perturbation per particle $\frac{1}{N}\sum_i \|(\Delta x_i, \Delta v_i)\|_2$, using the same statistical measures. The potential parameters are identical to Table 1 and the number of particles is $N = 100$. We observe from Fig. 1, that the probability of loosing additional polarization in the flock nonlinearly increases with the strength of the initial perturbation. Below a critical threshold of approximately 0.6, the flock will most likely drop polarization with an increasing probability of near zero values (e.g. unpredictable state). From Fig. 1(b) we observe that a perturbation-induced polarization of 0.6 is on average generated by a strong average perturbation in (x,v) of approximately 1.34 per particle, which on average is a perturbation of approximately 0.95 in position per particle. However, the mean nearest neighbour distance of the stationary state considered here is approximately 0.252. Hence the nonlinear stability observed in this experiment well exceeds the validity regime of Theorem 1, which is expected to hold for spatial perturbations that are small compared to the nearest neighbour distances.

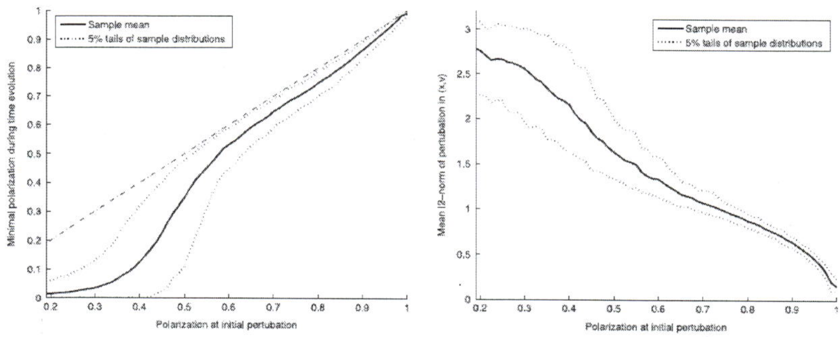

(a) Statistical comparison of initial polarization vs. minimal polarization in time evolution.

(b) Statistics of the mean-ℓ_2 perturbation per particle in (x, v) for the initial perturbations of (a).

Figure. 1 Numerical findings of a sequence of simulations of system (1) with perturbed initial data. (a): Histogram of initial polarization of the perturbed flock against the corresponding minimal polarization observed during the subsequent time evolution. (b): Supporting statistics of the initial polarization against the average ℓ_2-norm of perturbation per particle that induce the respective polarization change (–: arithmetic mean, — —: 5% tails of statistical distribution, ·–: identity function, Morse potential as in Table 1, $\alpha=1, \beta=5, \lambda=50, N=100$ particles, 25,000 simulations).

6. CONCLUSIONS & PERSPECTIVES

In this paper, we have shown that flock solutions are fully nonlinearly stable under small enough position–velocity perturbations, in the sense that a perturbed flock configuration will relax to another flock configuration in $\tilde{\mathcal{Z}}_F$. Stability is inherited from the first-order particle system to the second-order model. Hence the flock might slightly rotate its position, translate its centre of mass, change the orientation of its mean velocity, or any combination of the three, but it will remain a flock solution. This result has been established for general interaction potentials W satisfying assumptions (H1)–(H4). We numerically investigated the spectrum of the Jacobian $G(\hat{x})$ evaluated at a stationary state $\hat{x}$ for some commonly used potentials, and found them to possess the desired eigenvalue structure. Flocks for Morse-type potentials are linearly unstable in the co-moving frame second-order model [30], but nonlinearly asymptotically stable in the sense of our main result in Theorem 1. Finally, we investigated the nonlinear stability of flocks under uniformly distributed perturbations, and numerically validated the nonlinear stability of flocks in the sense of polarization.

Our findings are completely consistent with one's expectations from numerical experiments and, to the best of our knowledge, is the first of its kind for general flocks in second-order models. The task of theoretically proving nonlinear stability properties such as (H1)–(H4) of potentials and their steady solutions for the first-order aggregation equation remains largely open, and is inevitably complex given the extensive variety of stationary states (see e.g. [28]). An extension of the obtained result to three or higher dimensional space seems achievable, however with greater technical effort due to the increasing number of degrees of symmetry for the flock solution and the weakening of the interaction potential observed in [35]. The investigation of stability of other coherent patterns of swarming models such as rotating mills is another highly interesting open problem where the techniques used in this work clearly cannot be applied straightforwardly.

ACKNOWLEDGEMENTS

JAC acknowledges support from projects MTM2011-27739-C04-02, 2009-SGR-345 from Agència de Gestió d'Ajuts Universitaris i de Recerca-Generalitat de Catalunya, and the Royal Society through a Wolfson Research Merit Award. JAC, YH, and SM were supported by Engineering and Physical Sciences Research Council (UK) grant numberEP/K008404/1. Supporting materials can be obtained from the corresponding author upon request.

REFERENCES

1. J.K. Parrish, W.M. Hamner, Animal Groups in Three Dimensions: How Species Aggregate, Cambridge University Press, 1997.

2. I. Couzin, J. Krause, R. James, G.D. Ruxton, N.R. Franks, Collective memory and spatial sorting in animal group, J. Theoret. Biol. 218 (2002) 1–11.

3. I.D. Couzin, J. Krause, Self-organization and collective behavior of vertebrates, Adv. Study Behav. 32 (2003) 1–67.

4. F. Cucker, S. Smale, Emergent behavior in flocks, IEEE Trans. Automat. Control 52 (5) (2007) 852–862.

5. E. Cristiani, B. Piccoli, A. Tosin, Modeling self-organization in pedestrians and animal groups from macroscopic and microscopic viewpoints, in: Mathematical Modeling of Collective Behavior in Socio-Economic and Life Sciences, Springer, 2010, pp. 337–364.

6. S. Camazine, J.-L. Deneubourg, N.R. Franks, J. Sneyd, G. Theraulaz, E. Bonabeau, Self-organization in biological systems, in: Princeton Studies in Complexity, Princeton University Press, Princeton, NJ, 2003, Reprint of the 2001 original.

7. A. Mogilner, L. Edelstein-Keshet, A non-local model for a swarm, J. Math. Biol. 38 (1999) 534–570.

8. D.D. Holm, V. Putkaradze, Aggregation of finite-size particles with variable mobility, Phys. Rev. Lett. 95 (2005) 226106.

9. D.D. Holm, V. Putkaradze, Formation of clumps and patches in selfaggregation of finite-size particles, Physica D 220 (2) (2006) 183–196.

10. H. Levine, W.-J. Rappel, I. Cohen, Self-organization in systems of self-propelled particles, Phys. Rev. E 63 (2000) 017101.

11. M.R. D'Orsogna, Y.-L. Chuang, A.L. Bertozzi, L.S Chayes, Self-propelled particles with soft-core interactions: patterns, stability, and collapse, Phys. Rev. Lett. 96 (10) (2006) 104302.

12. Y.-L. Chuang, M.R. D'Orsogna, D. Marthaler, A.L. Bertozzi, L.S. Chayes, State transitions and the continuum limit for a 2D interacting, self-propelled particle system, Physica D 232 (1) (2007) 33–47.

13. J.A. Carrillo, M. Fornasier, G. Toscani, F. Vecil, Particle, kinetic, and hydrodynamic models of swarming, in: Mathematical Modeling of Collective Behavior in Socio-Economic and Life Sciences, Springer, 2010, pp. 297–336.

14. T. Vicsek, A. Czirók, E. Ben-Jacob, I. Cohen, O. Shochet, Novel type of phase transition in a system of self-driven particles, Phys. Rev. Lett. 75 (6) (1995) 1226.

15. P. Degond, S. Motsch, Continuum limit of self-driven particles with orientation interaction, Math. Models Methods Appl. Sci. 18 (supp01) (2008) 1193–1215.

16. J.A. Carrillo, M. Fornasier, J. Rosado, G. Toscani, Asymptotic flocking dynamics for the kinetic Cucker–Smale model, SIAM J. Math. Anal. 42 (1) (2010) 218–236.

17. I. Aoki, A simulation study on the schooling mechanism in fish, Bull. Japan Soc. Sci. Fish 48 (1982) 1081–1088.

18. A. Huth, C. Wissel, The simulation of fish schools in comparison with experimental data, Ecol. Model. 75/76 (1994) 135–145.

19. M. Ballerini, N. Cabibbo, R. Candelier, A. Cavagna, E. Cisbani, I. Giardina, V. Lecomte, A. Orlandi, G. Parisi, A. Procaccini, M. Viale, V. Zdravkovic, Interaction ruling animal collective behaviour depends on topological rather than metric distance: evidence from a field study, Proc. Natl. Acad. Sci. 105 (2008) 1232–1237.

20. A. Cavagna, A. Cimarelli, I. Giardina, G. Parisi, R. Santagati, F. Stefanini, R. Tavarone, From empirical data to inter-individual interactions: unveiling the rules of collective animal behavior, Math. Models Methods Appl. Sci. (2010) http://dx.doi.org/10.1142/S0218202510004660.

21. C.K. Hemelrijk, H. Hildenbrandt, Self-organized shape and frontal density of fish schools, Ethology 114 (2008) 245–254.

22. Alethea B.T. Barbaro, Kirk Taylor, Peterson F. Trethewey, Lamia Youseff, Björn Birnir, Discrete and continuous models of the dynamics of pelagic fish: application to the capelin, Math. Comput. Simul. 79 (12) (2009) 3397–3414.

23. H. Hildenbrandt, C. Carere, C.K. Hemelrijk, Self-organised complex aerial displays of thousands of starlings: a model, Behav. Ecol. 107 (21) (2010 1349–1359.

24. T. Kolokonikov, H. Sun, D. Uminsky, A. Bertozzi, Stability of ring patterns arising from 2d particle interactions, Phys. Rev. E 84 (1) (2011) 015203.

25. J.H. von Brecht, D. Uminsky, T. Kolokolnikov, A.L. Bertozzi, Predicting pattern formation in particle interactions, Math. Models Methods Appl. Sci. 22 (suppl. 1) (2012) 1140002, 31.

26. J.H. von Brecht, D. Uminsky, On soccer balls and linearized inverse statistical mechanics, J. Nonlinear Sci. 22 (6) (2012) 935–959.

27. D. Balagué, J.A. Carrillo, T. Laurent, G. Raoul, Nonlocal interactions by repulsive–attractive potentials: radial ins/stability, Physica D 260 (2013) 5–25.

28. D. Balagué, J.A. Carrillo, T. Laurent, G. Raoul, Dimensionality of local minimizers of the interaction energy, Arch. Rat. Mech. Anal. 209 (3) (2013) 1055–1088.

29. A.L. Bertozzi, J.H. von Brecht, H. Sun, T. Kolokolnikov, D. Uminsky, Ring patterns and their bifurcations in a nonlocal model of biological swarms, Commun. Math. Sci., in preparation.

30. G. Albi, D. Balagué, J.A. Carrillo, J. von Brecht, Stability analysis of flock and mill rings for 2nd order models in swarming, 2013, Preprint.

31. J.A. Carrillo, S. Martin, V. Panferov, A new interaction potential for swarming models, Physica D 260 (2013) 112–126.

32. J.A. Carrillo, Y. Huang, S. Martin, Explicit flock solutions for quasi-morse potentials, 2013, Preprint.

33. [33] M.W. Hirsch, C.C. Pugh, M. Shub, Invariant Manifolds, in: Number no. 583 in Lecture Notes in Mathematics, Springer-Verlag, 1977.

34. R.C. Fetecau, Y. Huang, T. Kolokolnikov, Swarm dynamics and equilibria for a nonlocal aggregation model, Nonlinearity 24 (10) (2011) 2681–2716.

35. F. Vecil, P. Lafitte, J. Rosado, A numerical study of attraction/repulsion collective behavior models: 3D particle analyses and 1D kinetic simulations, Physica D 260 (2013) 127–144.

7

CHAPTER

Resonance phenomenon for a Gelfand-type problem

Wenjing Chen, Juan Dávila

Departamento de Ingeniería Matemática and CMM, Universidad de Chile, Casilla 170 Correo 3, Santiago, Chile

ABSTRACT

We consider positive radially symmetric solutions of

$$-\Delta u = \lambda(e^u - 1), \quad \text{in } B, \qquad u = 0 \quad \text{on } \partial B,$$

where B is the unit ball in R^N, $N \geq 3$ and $\lambda > 0$ is a parameter. We establish infinite multiplicity of regular solutions for $3 \leq N \leq 9$ and some λ, and we obtain a bound for the Morse index and the number of solutions when $N \geq 10$.

Keywords: Singular solution; Multiplicity; Morse index; Resonance

1. INTRODUCTION

In this article, we are interested in the structure of the solution set of the boundary value problem

$$\begin{cases} -\Delta u = \lambda(e^u - 1), & u > 0 \quad \text{in } B; \\ u = 0 & \text{on } \partial B, \end{cases} \tag{1.1}$$

where B is the unit ball in $R^N, N \geq 3$ and $\lambda > 0$ is a parameter. Smooth solutions to(1.1) are radially symmetric and decreasing by the classical result of Gidas, Ni and Nirenberg [1].

Problem (1.1) is related to the following *Gelfand* problem:

$$\begin{cases} -\Delta u = \lambda e^u, & \text{in } B; \\ u = 0 & \text{on } \partial B. \end{cases} \tag{1.2}$$

Barenblatt [2] and Joseph and Lundgren [3], using phase-plane analysis, gave a complete description of the classical solutions to (1.2), which are again radially symmetric [1].

Proposition 1.1.

Assume $N \geq 1$, *then there exists* $\lambda^* = \lambda^*(N) > 0$, *such that*

- *for* $0 < \lambda < \lambda^*$, (1.2) *has the minimal solution* u_λ;
- *for* $\lambda = \lambda^*$, (1.2) *has a unique solution;*
- *for* $\lambda > \lambda^*$, (1.2) *has no solution (even in the weak sense).*

Moreover, we have the following.

(a) *if* $N = 1, 2$, *then for* $0 < \lambda < \lambda^*$, *there are exactly two solutions to* (1.2), *one of them is the minimal solution* u_λ. *The other one, denoted by* U_λ, *has Morse index* 1.

(b) *If* $3 \leq N \leq 9$, *then* $\lambda^* > 2(N-2)$. *For* $0 < \lambda < \lambda^*, \lambda \neq 2(N-2)$, (1.2) *has finitely many solutions; for* $\lambda = 2(N-2)$, (1.2) *has infinitely many solutions; for* λ *close to* $2(N-2)$, (1.2) *has a large number of solutions that converge to* $-2\log|x|$.

(c) *If* $N \geq 10$, *then* $\lambda^* = 2(N-2)$ *and* $u_* = -2\log|x|$. *Moreover* (1.2) *has a unique minimal solution* u_λ *for each* $\lambda \in (0, \lambda^*)$.

Nagasaki and Suzuki [4] classified the solutions of (1.2) according to their Morse index. In a few words, the family of regular solutions of (1.2) can be described as a curve $(u(s), \lambda(s))$ with $s \in [0, \infty)$, such that $(u(s), \lambda(s)) \to (0,0)$ as $s \to 0$ and $(u(s), \lambda(s)) \to (u_\sigma, \lambda_\sigma)$ as $s \to \infty$, where $u_\sigma(r) = -2\log(r), \lambda_\sigma = 2(N-2)$ is a singular solution of (1.2). In dimensions $3 \leq N \leq 9, \lambda(s)$ oscillates around $2(N-2)$ as $s \to \infty$ and the Morse index of $u(s)$ increases by one in each oscillation. In dimensions $N \geq 10, \lambda(s)$ is monotone, $u(s)$ is monotone and is stable for each s. We refer the reader to the book of L. Dupaigne [5] for further references on problem (1.2). Moreover, Berchio, Gazzola and Pierotti in [6] studied Gelfand type elliptic problems under Steklov boundary conditions.

A problem analogous to (1.1) is

$$\begin{cases} -\Delta u = u^p + \lambda u, & u > 0 \quad \text{in } B; \\ u = 0 & \text{on } \partial B \end{cases} \tag{1.3}$$

where $p > 1$ and $\lambda > 0$ is a parameter. According to classical bifurcation theory [7], the point $(\mu_1, 0)$ is a bifurcation point from which emanates an unbounded branch C of solutions of (1.3), where μ_1 is the first eigenvalue of the negative Laplacian operator under Dirichlet boundary condition in B.

•If $p < \frac{N+2}{N-2} (N \geq 3)$, for $\lambda < \mu_1$, there is a positive solution of (1.3) by a standard constrained minimization procedure involving compactness of the Sobolev embedding. Moreover, by Pohozaev's identity [8], problem (1.3) has no solutions for $\lambda \leq 0$ whenever $p \geq \frac{N+2}{N-2}$.

•If $p = \frac{N+2}{N-2}$, which is the classical Brezis–Nirenberg problem [9], problem (1.3) has a solution for $0 < \lambda < \mu_1$ if $N \geq 4$, and for $\frac{1}{4}\mu_1 < \lambda < \mu_1$ if $N = 3$.

•If $p > \frac{N+2}{N-2}$, Dolbeault and Flores found that if $p > \frac{N+2}{N-2}$, and $p < \frac{N-2\sqrt{N-1}}{N-2\sqrt{N-1}-4}$ or $N \leq 10$, then there is a unique number $\lambda_* > 0$, such that for λ close to λ_*, a large number of classical solutions of (1.3) exist. In particular, there are infinitely many classical solutions for $\lambda = \lambda_*$. Recently, Guo and Wei in [10] showed that the structure of the branch C changes for $p \geq p_c$ and $\frac{N+2}{N-2} < p < p_c$, where $p_c = \frac{(N-2)^2-4N+8\sqrt{N-1}}{(N-2)(N-10)}$ if $N \geq 11$; and $p_c = \infty$ if $2 \leq N \leq 10$. Moreover, they established that for $\frac{N+2}{N-2} < p < p_c$, C turns infinitely many times around $\lambda_* \in (0, \mu_1)$. For $p \geq p_c$, all solutions have a finite Morse index, and for $N \geq 12$ and $p > p_c$ sufficiently large all solutions have exactly Morse index one.

This paper is devoted to the study of the structure of solutions to problem (1.1). We start with some general remarks. First, classical solutions of (1.1) can exist only for λ in some interval.

Proposition 1.2.

Let μ_1 *be the first eigenvalue of the* $-\Delta$ *under Dirichlet boundary condition in* B. *Then there exists* $\lambda_0 > 0$, *such that a necessary condition for existence of classical solutions to problem* (1.1) *is* $\lambda \in (\lambda_0, \mu_1)$.

See a proof in the Appendix. By classical bifurcation theory [11] and [7] we have that $(\mu_1, 0)$ is a bifurcation point of solutions to (1.1). Both observations are also valid if we replace the ball by a bounded smooth domain (star shaped in the case ofProposition 1.2).

We are interested also in weak solutions, allowing for possible singularities.

Definition 1.3.

We say that $u \in H_0^1(B)$ is a weak solution of (1.1) if $e^u \in L^1(B)$ and

$$\int_B \nabla u \nabla \varphi = \lambda \int_B (e^u - 1)\varphi \quad \text{for all } \varphi \in C_0^\infty(B). \tag{1.4}$$

We say that a weak solution u of (1.1) is regular (resp., singular) if $u \in L^{\infty}(B)$ (resp.,$u \notin L^{\infty}(B)$).

We say that a radial weak solution u of (1.1) is a weakly singular solution if it is singular and $\lim_{r\to 0} ru'(r)$ exists.

We first study singular solutions to (1.1).

Theorem 1.4.

Assume $N \geq 3$. *Let* $\lambda > 0$ *and suppose that* $u \in C^2(B\setminus\{0\})$, $u \geq 0$ *is a radial solution of*

$$-\Delta u = \lambda(e^u - 1) \quad in\ B\setminus\{0\}. \tag{1.5}$$

Then either

(a) u *can be extended as a function in* $C^{\infty}(B)$ *and* (1.5) *holds in* B,

or

(b) u *is singular at* $r = 0$ *and satisfies*

$$\lim_{r\to 0}(u(r) + 2\log r) = \log\frac{2(N-2)}{\lambda},$$

$$\lim_{r\to 0} ru'(r) = -2.$$

As a consequence, u *is a radial singular weak solution to* (1.1) *if and only if* u *is a weakly singular solution.*

Theorem 1.5.

For $N \geq 3$, *there exists a unique* $\lambda_* > 0$, *such that* (1.1) *admits a radial singular solution for* $\lambda = \lambda_*$, *and the radial singular solution is unique.*

By Theorem 1.4 the singular solution is weakly singular.

Next, we consider the question of multiplicity of solutions to (1.1).

Theorem 1.6.

If $3 \leq N \leq 9$, *then problem* (1.1) *has infinitely many regular radial solutions for* $\lambda = \lambda_*$. *For* $\lambda \neq \lambda_*$ *but close to* λ_*, *there is a large number of regular radial solutions for* (1.1).

For a weak solution (λ, u) of (1.1) we define the Morse index of u as the largest dimension k of a subspace $Y \subset C_c^\infty(B)$ such that

$$Q_u(\varphi) = \int_B |\nabla\varphi|^2 - \lambda e^u \varphi^2 < 0 \quad \forall \varphi \in Y \setminus \{0\}.$$

If u is a regular solution this is the number of negative eigenvalues, counting multiplicity, of the operator $-\Delta - \lambda e^u$. By Theorem 3 of Dancer and Farina [12], if $3 \leq N \leq 9$, for a sequence of solutions $(\lambda_n, u_{\lambda n})$ to (1.1) with $\|u_n\|_{L^\infty(B)} \to \infty$ as $n \to \infty$, then the Morse index of $u_{\lambda n}$ goes to infinity as $n \to \infty$.

Theorem 1.7.

Assume $N \geq 10$. *Then there exists* $K < \infty$ *such that the Morse index of any radial solution* (λ, u_λ) *of* (1.1) *(regular or singular) is bounded by* K. *The number of intersections of any regular solution and the radial singular solution is uniformly bounded by* $2K+1$. *Moreover, for each* $\lambda \in (\lambda_0, \mu_1)$, *the number of regular solutions to* (1.1) *is bounded by* $(K+1)^2$.

A natural conjecture for $N \geq 10$, which is observed in numerical calculations, is that the Morse index of any radial solution of (1.1) (regular or singular) is 1, the number of intersections of any regular solution and the radial singular solution is 1, and that for each $\lambda \in (\lambda_*, \mu_1)$ there is a unique solution.

To obtain multiplicity of solutions to problem (1.1) we use geometric theory of dynamical systems in three-dimensional phase space, which was applied in [13], and subsequently in [14], [15] and [16]. There are some analogies between the results and techniques of this work and [17], [18], [19], [20] and [21] on fourth order problems involving the exponential nonlinearity. In Section 2 we give some preliminaries. In Section 3 we prove Theorem 1.4, namely that radial solutions either are regular or weakly singular. Theorem 1.5, which is about the existence and uniqueness of a singular solution is proved in Section 4.

In Section 5 we prove Theorem 1.6 on the multiplicity of solutions in dimensions $3 \leq N \leq 9$. In Section 6 we analyze the Morse index of solutions to problem(1.1), give the structure of the branch of solutions to (1.1), and prove Theorem 1.7. Finally, we give the proof of Proposition 1.2 in the Appendix.

2. PRELIMINARY RESULTS

Let u satisfy (1.1) and make the change of variables

$$v(t) = u(r) \quad \text{with} \, r = e^t, \, \text{for} \, t \in (-\infty, 0). \tag{2.1}$$

Then problem (1.1) becomes

$$\begin{cases} -v''(t) + (2-N)v'(t) = \lambda e^{2t}(e^{v(t)} - 1), & t \in (-\infty, 0) \\ v(0) = 0, & \lim_{t \to -\infty} e^{-t} v'(t) = 0. \end{cases} \tag{2.2}$$

Define

$$\begin{cases} v_1(t) = \dfrac{\lambda}{2(N-2)} e^{v(t)+2t}, \\ v_2(t) = v'(t), \\ v_3(t) = \lambda e^{2t}. \end{cases} \tag{2.3}$$

We find that (v_1, v_2, v_3) satisfies the following differential system

$$\begin{cases} v_1' = v_1(v_2 + 2), \\ v_2' = -2(N-2)v_1 - (N-2)v_2 + v_3, \\ v_3' = 2v_3, \end{cases} \tag{2.4}$$

with the condition

$$v_3(0) = 2(N-2)v_1(0). \tag{2.5}$$

System (2.4) has two stationary points

$$P_1 = (0,0,0) \text{ and } P_2 = (1,-2,0).$$

The linearization of (2.4) around P_1 is given by $X' = M_1X$, with

$$M_1 = \begin{bmatrix} 2 & 0 & 0 \\ -2(N-2) & 2-N & 1 \\ 0 & 0 & 2 \end{bmatrix}.$$

The eigenvalues of M_1 are $\tilde{\nu}_1 = \tilde{\nu}_2 = 2, \tilde{\nu}_3 = 2-N$. Thus for $N \geq 3, P_1 = (0,0,0)$ is a hyperbolic point, which has a 2-dimensional unstable manifold $W^u(P_1)$ and a 1-dimensional stable manifold $W^s(P_1)$.

The linearization of (2.4) around P_2 is given by $X' = M_2X$, with

$$M_2 = \begin{bmatrix} 0 & 1 & 0 \\ -2(N-2) & 2-N & 1 \\ 0 & 0 & 2 \end{bmatrix}. \tag{2.6}$$

The eigenvalues of M_2 are given by

$$\nu_1 = 2, \qquad \nu_{2,3} = \frac{(2-N) \pm \sqrt{(N-2)(N-10)}}{2}. \tag{2.7}$$

For $3 \leq N \leq 9, \nu_2$ and ν_3 are complex conjugates and $\mathrm{Re}(\nu_2) = \mathrm{Re}(\nu_3) = \frac{2-N}{2} < 0$. For $N \geq 10$, all the eigenvalues are real and $\nu_1 > 0, \nu_2 < 0, \nu_3 < 0$. Thus for all $N \geq 3, P_2 = (1,-2,0)$ is a hyperbolic point, which has a 1-dimensional unstable manifold $W^u(P_2)$ and a 2-

dimensional stable manifold $W^s(P_2)$. Actually $W^s(P_2)$ is contained in the plane $\{v_3=0\}$, which is invariant for (2.4).

Also we note that solutions of system (2.4) restricted to $\{v_3=0\}$ are related to radial solutions of the equation

$$-\Delta u = \lambda e^u \tag{2.8}$$

by exactly the same change of variables (2.1) and the first two equations in (2.3). This yields immediately a heteroclinic connection from P_1 to P_2, which is associated to the unique radial solution of (2.8) with $\lambda=2(N-2)$ and initial condition $u(0)=u'(0)=0$.

Proposition 2.1.

For $N\geq 3$, *system* (2.4) *has a heteroclinic orbit from* P_1 *to* P_2, *which is contained in the plane* $\{v_3=0\}$.

Thanks to a result of Belickiĭ [22], we have the following lemma.

Lemma 2.2.

The system (2.4) *is* C^1*-conjugate to its linearization around* $P_2=(1,-2,0)$.

Proof.

We just need to check that none of the following relations

$$\mathrm{Re}(\nu_i) = \mathrm{Re}(\nu_j) + \mathrm{Re}(\nu_k), \tag{2.9}$$

holds for different indices $i,j,k\in\{1,2,3\}$ such that $\mathrm{Re}(\nu_j)<0$ and $\mathrm{Re}(\nu_k)>0$, where ν_1,ν_2,ν_3 are corresponding eigenvalues of M_2. It is easy to check this by calculation for $N\geq 3$. □

Lemma 2.3.

Let $v^{(1)},v^{(2)},v^{(3)}$ *be the eigenvectors of* M_2 *associated to* ν_1,ν_2,ν_3. *Then* $v^{(k)}=(1,\nu_k,\nu_k(\nu_k-(2-N))+2(N-2))$ *and* $v^{(1)}$ *is always real;*

for $3 \leq N \leq 9, v^{(2)}, v^{(3)}$ *are complex conjugates. In particular the components of* $v^{(1)} = (1, 2, 4(N-1))$ *are positive.*

Proof.

By direct calculation, $v^{(k)} = (1, \nu_k, \nu_k(\nu_k - (2-N)) + 2(N-2))$ is an eigenvector associated to ν_k. □

3. CHARACTERIZATION OF WEAKLY SINGULAR SOLUTIONS

In this section our aim is to prove Theorem 1.4. We assume that $u \in C^2(0,1)$, $u \geq 0$ satisfies

$$-\Delta u = 2(N-2)(e^u - 1) \quad \text{in } (0,1), \tag{3.1}$$

where we assume, by using a scaling, that $\lambda = 2(N-2)$. The scaling changes the length of the interval where the solution is defined, but this is not relevant for the next arguments, so we assume that the interval is (0, 1).

Define $v(t) = u(e^t), w(t) = v(t) + 2t$ for $t \leq 0$. Then w satisfies

$$-w''(t) + (2-N)w'(t) = 2(N-2)\left(e^{w(t)} - e^{2t} - 1\right) \quad \text{for all } t \leq 0. \tag{3.2}$$

We also let v_1, v_2, v_3 be defined in (2.3).

By similar arguments as in [20], we have the following results.

Lemma 3.1.

One has

$$\liminf_{t\to-\infty} w(t) \le 0. \tag{3.3}$$

Proof.

We follow [23]. Let $L := \liminf_{t\to-\infty} w(t)$ and suppose by contradiction that $L>0$. Then there exists $T_0>0$, such that $w(t)\ge L/2$ for all $t\le -T_0$. Let ϕ be a smooth cut-off function in R such
that $0\le\phi(t)\le 1$, $\phi(t)=0$ for $t\le-(T_0+3)$ and $t\ge -T_0$; $\phi(t)=1$ for $t\in[-(T_0+2),-(T_0+1)]$, and for $i=1,2$

$$\int_{-(T_0+3)}^{-T_0} \frac{(\phi^{(i)})^2}{\phi}\, dt := c_i < +\infty.$$

Let $\tau>1$ and $\phi_\tau(t) = \phi(\frac{t}{\tau})$. Multiplying (3.2) by ϕ_τ and integrating, we get

$$\int_{-(T_0+3)\tau}^{-T_0\tau} (e^{w(t)} - 1)\phi_\tau\, dt = \sum_{i=1}^{2} a_i \int_{-(T_0+3)\tau}^{-T_0\tau} w\phi_\tau^{(i)}\, dt + \int_{-(T_0+3)\tau}^{-T_0\tau} e^{2t}\phi_\tau\, dt, \tag{3.4}$$

where $a_1 = \frac{1}{2}, a_2 = -\frac{1}{2(N-2)}$. Using Young's inequality with $\varepsilon_1>0$ to be fixed later on, we have

$$\begin{aligned} \left|\int_{-(T_0+3)\tau}^{-T_0\tau} w\phi_\tau^{(i)}\, dt\right| &\le \varepsilon_1 \int_{-(T_0+3)\tau}^{-T_0\tau} w^2\phi_\tau\, dt + C_{\varepsilon_1}\int_{-(T_0+3)\tau}^{-T_0\tau} \frac{(\phi_\tau^{(i)})^2}{\phi_\tau}\, dt \\ &\le \varepsilon_1 \int_{-(T_0+3)\tau}^{-T_0\tau} w^2\phi_\tau\, dt + C_{\varepsilon_1} c_i \tau^{1-2i}. \end{aligned} \tag{3.5}$$

We also have

$$\int_{-(T_0+3)\tau}^{-T_0\tau} e^{2t}\phi_\tau \, dt \leq \frac{1}{2}e^{-2T_0\tau}. \tag{3.6}$$

From (3.4), (3.5) and (3.6) we get

$$\int_{-(T_0+3)\tau}^{-T_0\tau} \left[e^{w(t)} - 1 - \varepsilon_1 K w(t)^2\right] \phi_\tau \, dt \leq C_{\varepsilon_1} K \max_{i=1,2} c_i \tau^{1-2i} + \frac{1}{2}e^{-2T_0\tau}$$

with $K=|a_1|+|a_2|$. Since $w(t) \geq L/2 > 0$ for all $t \leq -T_0$, we can choose $\varepsilon_1 > 0$ small, such that $e^{w(t)} - 1 - \varepsilon_1 K w(t)^2 \geq \varrho$ for $t \leq -T_0$, where $\varrho > 0$ is fixed. Then

$$\varrho\tau \leq \int_{-(T_0+3)\tau}^{-T_0\tau} \left[e^{w(t)} - 1 - \varepsilon_1 K w(t)^2\right] \phi_\tau \, dt \leq C_{\varepsilon_1} K \max_{i=1,2} c_i \tau^{1-2i} + \frac{1}{2}e^{-2T_0\tau},$$

which is impossible for $\tau > 1$ large.

Lemma 3.2.

We have

$$\limsup_{t \to -\infty} w(t) < +\infty.$$

Proof.

Assume by contradiction that $\limsup_{t\to-\infty} w(t) = +\infty$. Then there is a sequence $t_k \to -\infty$ such that $w(t_k) \to +\infty$. Furthermore we can assume that for all $k \geq 1$ we have $t_{k+1} + \log 2 < t_k$, $w(t_{k+1}) \geq w(t_k)$.

Set $M_k = w(t_k)$, $r_k = e^{t_k}$ and $\rho_k = \frac{r_{k+1}}{r_k}$. Note that $0 < \rho_k < \frac{1}{2}$. Let $\eta_k(r) = \frac{N-2}{N} r_k^2 (1 - r^2)$ so that it satisfies

$$-\Delta\eta_k = 2(N-2)r_k^2 \ \text{ in } B, \qquad \eta_k = 0 \ \text{ on } \partial B.$$

Define

$$u_k(r) = u(rr_k) - M_k + 2\log(r_k) + \eta_k(r).$$

Then we have

$$-\Delta u_k(r) = 2(N-2)r_k^2 e^{u(r_k r)} = 2(N-2)e^{M_k-\eta_k(r)}e^{u_k(r)}, \quad \text{for } 0 < r < r_k^{-1}.$$

Since η_k is bounded from above,

$$-\Delta u_k \geq C_0 e^{M_k} e^{u_k} \quad \forall 0 < r < r_k^{-1}, \tag{3.7}$$

for some $C_0 > 0$ independent of k. Also note that

$$u_k(1) = u(r_k) - M_k + 2t_k = 0,$$
$$u_k(\rho_k) = M_{k+1} - M_k + 2(t_k - t_{k+1}) + \eta_k(\rho_k) \geq 0.$$

Let $\lambda_{1,k}$ be the first eigenvalue for $-\Delta$ with Dirichlet boundary condition in the annulus $B \setminus B_{\rho k}$ and $\phi_k > 0$ be the corresponding eigenfunction, that is,

$$\begin{cases} -\Delta\phi_k = \lambda_{1,k}\phi_k, \quad \phi_k > 0 & \text{in } B\setminus B_{\rho_k}; \\ \phi_k = 0; & \text{on } \partial\left(B\setminus B_{\rho_k}\right), \end{cases}$$

normalized so that $\|\phi_k\|_{L^\infty(B)} = 1$. Multiplying (3.7) by ϕ_k and integrating in $B \setminus B_{\rho k}$, we get

$$C_0 e^{M_k} \int_{B \setminus B_{\rho_k}} e^{u_k} \phi_k \, dx \leq \int_{\partial(B \setminus B_{\rho_k})} \frac{\partial \phi_k}{\partial \nu} u_k \, d\sigma + \lambda_{1,k} \int_{B \setminus B_{\rho_k}} u_k \phi_k \, dx.$$

But $u_k \geq 0$ and $\frac{\partial \phi_k}{\partial \nu} \leq 0$ on $\partial(B \setminus B_{\rho k})$ so that

$$C_0 e^{M_k} \int_{B \setminus B_{\rho_k}} e^{u_k} \phi_k \, dx \leq \lambda_{1,k} \int_{B \setminus B_{\rho_k}} u_k \phi_k \, dx.$$

Now using the inequality $e^u \geq u$, it yields that

$$C_0 e^{M_k} \leq \lambda_{1,k}.$$

However, since the annulus $B \setminus B_{\rho k}$ has a width that does not converge to zero, $\lambda_{1,k}$ remains uniformly bounded. It follows that M_k is bounded as $k \to \infty$, which is a contradiction.

Lemma 3.3.

For $i=0,1,2$, *we have*

$$|w^{(i)}(t)| \leq C(1+|t|) \quad \text{for all } t \leq 0, \tag{3.8}$$

and for all $i=1,2,3$

$$|v_i(t)| \leq C(1+|t|) \quad \text{for all } t \leq 0. \tag{3.9}$$

Proof.

Since $u \geq 0$ and w is bounded above, we have $|w(t)| \leq C(1+|t|)$. Moreover, by Eq.(3.2), and interpolation inequalities such as in Chapter 6 of [24], we get that for any $t \leq -1$ and $i = 1,2$

$$\begin{aligned} |w^{(i)}(t)| &\leq C \sup_{[t-1,t+1]} \left(|w| + 2(N-2)|e^{w} - e^{2t} - 1|\right) \\ &\leq C \sup_{[t-1,t+1]} \left(|w| + 2(N-2)|e^{w} - 1|\right). \end{aligned}$$

Since w is bounded above, the second term in the supremum is bounded. Then(3.8) and (3.9) follow from the bound of w. □

Lemma 3.4.

For $i = 1,2,3$

$$|v_i(t)| \leq C \quad \text{for all } t \leq 0, \tag{3.10}$$

for $i = 1,2$

$$|w^{(i)}(t)| \leq C \quad \text{for all } t \leq 0. \tag{3.11}$$

Proof.

It is direct that v_3 is bounded for all $t \leq 0$. Since $v_1(t) = e^{w(t)}$ (recall the change of variables (2.3) and that we assume $\lambda = 2(N-2)$) and w is bounded above, we have$v_1(t)$ is bounded as $t \to -\infty$. Next we prove that v_2 is bounded for all $t \leq 0$.

Integrating the following equation

$$\frac{d}{ds}\left(v_2(s)e^{(N-2)s}\right) = [-2(N-2)v_1(s) + v_3(s)]\, e^{(N-2)s}$$

in $[t,t_0]$ with $t\le t_0\le 0$, we get

$$v_2(t)=e^{-(N-2)t}\left(v_2(t_0)e^{(N-2)t_0}+2(N-2)\int_t^{t_0}e^{(N-2)s}v_1(s)\,ds-\frac{2(N-2)}{N}(e^{Nt_0}-e^{Nt})\right).$$

Since v_1 is bounded, the integral $\int_{-\infty}^{t_0}e^{(N-2)s}v_1(s)\,ds$ exists. If

$$\frac{2(N-2)}{N}e^{Nt_0}-2(N-2)\int_{-\infty}^{t_0}e^{(N-2)s}v_1(s)\,ds\neq v_2(t_0)e^{(N-2)t_0},$$

we deduce that $|v_2(t)|$ grows exponentially as $t\to-\infty$, which contradicts (3.9). Therefore we get

$$v_2(t_0)=-2(N-2)e^{-(N-2)t_0}\int_{-\infty}^{t_0}e^{(N-2)s}v_1(s)\,ds+\frac{2(N-2)}{N}e^{2t_0}\quad\forall\, t_0\le 0. \tag{3.12}$$

It follows that $|v_2(t)|\le C$ for all $t\le 0$, because v_1 is bounded.

Finally, the relations

$$w'(t)=v_2+2,\qquad w''(t)=-2(N-2)v_1+(2-N)v_2+v_3,$$

imply (3.11).

Proof of Theorem 1.4.

The statements in the theorem are consequence of the following properties, that we will prove next.

(i)If $\lim\inf_{t\to-\infty}w(t)=-\infty$, then $w(t)\to-\infty$, $v_i(t)\to 0$ as $t\to-\infty$ for $i=1,2,3$, andu is a regular solution.

(ii) If $\lim\inf_{t\to-\infty}w(t)>-\infty$, then $w(t)\to 0,(v_1,v_2,v_3)\to P_2$ as $t\to-\infty$, and u is a weakly singular solution.

To prove these claims it is useful to define

$$E(t) = \frac{1}{2}(w'(t))^2 + 2(N-2)(e^{w(t)} - w(t)) - (N-2)C_1 e^{2t},$$

where $C_1 > 0$ is a constant such that $|w'(t)| \le C_1$ for all $t \le 0$. This constant exists thanks to Lemma 3.4. Let us compute

$$E'(t) = (w(t)'' + 2(N-2)(e^{w(t)} - 1))w(t)' - 2(N-2)C_1 e^{2t}$$

for $t \le 0$. Using Eq. (3.2) we get

$$E'(t) = -(N-2)w'(t)^2 + 2(N-2)e^{2t}(w'(t) - C_1) \le 0. \quad (3.13)$$

Let us prove (i) and so we assume $\lim\inf_{t\to-\infty} w(t) = -\infty$. First, we show that $w(t) \to -\infty$ as $t \to -\infty$. By contradiction, we assume that $w(t)$ does not tend to $-\infty$ as $t \to -\infty$. Then we can find sequences $s_k \to -\infty, \tau_k \to -\infty$, such that $s_k > \tau_k$,

$$w(s_k) \to -\infty, \quad w(\tau_k) \text{ is bounded.}$$

But then $E(\tau_k)$ is bounded and $E(s_k) \to \infty$ as $k \to \infty$. However, by (3.13),$E(s_k) \le E(\tau_k)$, which is a contradiction.

Now, since $w(t) \to -\infty$ as $t \to -\infty$, we can easily deduce $v_1(t) \to 0$ as $t \to -\infty$. Using formula (3.12), we obtain $v_2(t) \to 0$ as $t \to -\infty$. Therefore $\lim_{t\to-\infty} V(t) = P_1$.

Since $v_2(t) \to 0$ as $t \to -\infty$, we have $\lim_{r\to 0} ru'(r) = 0$. Then for any $\epsilon > 0$, there exists $r_0 > 0$ such that for any $0 < r < r_0$, we have $|ru'(r)| < \epsilon$. Integrating from r to r_0 in this inequality, for any $0 < r < r_0$ we obtain

$$0 \leq u(r) \leq -\epsilon \ln r + C, \qquad e^{u(r)} \leq Cr^{-\epsilon}, \tag{3.14}$$

for some $C>0$.

We can then get that $u'(r)$ is bounded for $r>0$ small enough. In fact, Eq. (1.1) can be written as

$$-\left(s^{N-1}u'(s)\right)' = \lambda s^{N-1}\left(e^{u(s)} - 1\right).$$

Integrating above equation from δ to r with $(\delta,r)\subset(0,r_0)$ and using (3.14), letting $\delta\to 0$, we have

$$|u'(r)| \leq Cr^{1-N}\int_0^r s^{N-1}\left(s^{-\epsilon} - 1\right) ds \leq C$$

for $0<r<r_0$. From the boundedness of u' near $r=0$ we also get that u is bounded near $r=0$. This shows that u is regular.

We prove now (ii), so we assume that $\lim \inf_{t\to-\infty} w(t) > -\infty$. Since w is bounded above by Lemma 3.2, we have w is bounded. By Lemma 3.4, the derivatives of w are bounded, then we get that $E(t)$ is bounded as $t\to-\infty$. From the boundedness of E together with the boundedness of the derivatives of w and (3.13), we deduce that

$$\int_{-\infty}^{0} w'(t)^2\, dt < +\infty. \tag{3.15}$$

Set $\psi_T(t)=w'(t+T)$, then we get that

$$\psi_T \to 0 \quad \text{in } L^2(0,1) \text{ as } T \to -\infty.$$

Moreover, ψ_T satisfies the equation

$$-\psi_T''(t) + (2-N)\psi_T'(t) = 2(N-2)e^{w(T+t)}\psi_T(t) - 4(N-2)e^{2(T+t)}.$$

Using regularity theory, we have $\psi_T(\frac{1}{2}) \to 0$ and $\psi_T'(\frac{1}{2}) \to 0$ as $T \to -\infty$. Thus we obtain that $w'(t) \to 0$ as $t \to -\infty$ and similarly $w''(t) \to 0$ as $\to -\infty$. This implies that $\lim_{t\to-\infty} v'(t) = -2$. Since $v'(t) = u'(e^t)e^t$ we see that u is a weakly singular solution by the definition. We get in addition that $(v_1, v_2, v_3) \to (1, -2, 0)$ as $t \to -\infty$. That is, $\lim_{t\to-\infty} V(t) = P_2$. □

A direct corollary of the proof of Theorem 1.4 is the following.

Corollary 3.5.

Let u *be a radial singular solution to* (1.1) *and let* $V(t) = (v_1(t), v_2(t), v_3(t))$ *be the corresponding trajectory to* (2.4). *Then* $\lim_{t\to-\infty} V(t) = P_2 = (1, -2, 0)$.

As a consequence of Theorem 1.4 and Corollary 3.5, we have the following.

Corollary 3.6.

For u *a radial solution of* (1.1) *we have:*

(a) u *is regular if and only if* $\lim_{t\to-\infty} V(t) = P_1$;

(b) u *is singular if and only if* $\lim_{t\to-\infty} V(t) = P_2$.

4. THE UNSTABLE MANIFOLD AT P_2

In this section, we study the unstable manifold of P_2 and prove Theorem 1.5. First we have the following result.

Proposition 4.1.

Let $V(t)=(v_1(t),v_2(t),v_3(t)):(-\infty,T)\to R^3$ *be the trajectory in* $W^u(P_2)$ *such that* $v_3'(t)>0$ *as* $t\to-\infty$, *where* T *is the maximal time of existence. Then there exists some* $t<T$ *such that* $v_3(t)\geq 2(N-2)v_1(t)$.

Proof.

First we observe that this trajectory satisfies

$$v_1'(t)>0, \qquad v_2'(t)>0, \qquad v_3'(t)>0$$

for t close to $-\infty$ since the tangent vector to this trajectory becomes parallel to $(1,2,4(N-1))$ as it approaches P_2.

Let $z(t)=v_3(t)-2(N-2)v_1(t)$ and by contradiction we assume that

$$z(t)<0 \quad \text{for} \, \forall t\in(-\infty,T). \tag{4.1}$$

First, we remark that

$$v_2(t)<0 \quad \text{for} \, \forall t\in(-\infty,T). \tag{4.2}$$

To prove this, let us suppose it fails, and so there is the first time $t_0\in(-\infty,T)$, such that $v_2(t_0)=0$. Since $\lim_{t\to-\infty}v_2(t)=-2$ we must have $v_2'(t_0)\geq 0$. But writing the second equation in (2.4) as

$$v_2'(t)=z(t)-(N-2)v_2(t)$$

we would get $z(t_0)\geq 0$, a contradiction with (4.1).

Using (2.4) and $v_2(t)<0$ for all $t<T$ we can assert that the solution is defined for all t, that is $T=+\infty$. Indeed, the first equation in (2.4) yields

$$v_1(t) = v_1(t_0)e^{\int_{t_0}^{t}(2+v_2(s))\,ds} \tag{4.3}$$

Since $v_2(t)<0$ we see that $v_1(t)$ cannot blow up as $t\to T$, if T were finite. Also v_3 cannot blow up. This and the linearity of the second equation in (2.4) yield that $T=+\infty$.

Now, let us establish that

$$v_1(t) > 0 \quad \text{for } \forall t \in (-\infty, +\infty). \tag{4.4}$$

In fact, this is valid for t near $-\infty$ since $v_1(t)\to 1$ as $t\to-\infty$. If inequality (4.4) does not hold, then $v_1(t_0)=0$ for some t_0, and it follows from (4.3) that $v_1(t)=0$ for all t, a contradiction.

Next, we prove that

$$\limsup_{t\to+\infty} v_2(t) = 0. \tag{4.5}$$

Indeed, suppose not, we assume that there is a small number $\delta>0$ such that $v_2(t)<-\delta<0$ for all t. From the first equation in (2.4), we then get $v_1'(t) < (2-\delta)v_1(t)$, so we have $v_1(t)<v_1(0)e^{(2-\delta)t}$ for all $t>0$. But by the third equation in (2.4), we have $v_3(t)=v_3(0)e^{2t}$. Hence $z(t)=v_3(0)e^{2t}-2(N-2)v_1(0)e^{(2-\delta)t}\geq 0$ for some $t>0$, which contradicts assumption (4.1).

From (4.2) and (4.5), there exists a sequence (t_k) with $t_k\to+\infty$ as $k\to+\infty$, such that

$$v_2'(t_k) > 0, \quad \text{and} \quad v_2(t_k) \to 0 \quad \text{as } k \to +\infty.$$

Moreover, by the second equation in (2.4) we have $0 > z(t_k) = v_2'(t_k) + (N-2)v_2(t_k) > (N-2)v_2(t_k)$. Therefore,

$$z(t_k) \to 0 \quad \text{as } k \to +\infty. \tag{4.6}$$

From (2.4), we have $z'(t)-2z(t)=-2(N-2)v_1(t)v_2(t)$. Multiplying by e^{-2t} and integrating from t to t_k, we get

$$z(t_k) = e^{2(t_k-t)}\left(z(t) - 2(N-2)e^{2t}\int_t^{t_k} e^{-2s}v_1(s)v_2(s)ds\right). \tag{4.7}$$

From (4.2), (4.4), (4.6) and (4.7) we have that

$$\int_t^{+\infty} e^{-2s}v_1(s)|v_2(s)|ds < +\infty \quad \text{for any } t < +\infty. \tag{4.8}$$

Note that $v_1(t) = \frac{v_3(t)-z(t)}{2(N-2)}$ and hence

$$z'(t) - 2z(t) = (z(t) - v_3(t))v_2(t).$$

Multiplying by e^{-2t} and integrating from 0 to t_k, we find

$$z(t_k) = e^{2t_k}\left(z(0) + \int_0^{t_k} e^{-2s}z(s)v_2(s)ds - \int_0^{t_k} e^{-2s}v_2(s)v_3(s)ds\right).$$

Since $z(0) < 0$, $\int_0^{t_k} e^{-2s}z(s)v_2(s)ds$ and $-\int_0^{t_k} e^{-2s}v_2(s)v_3(s)ds$ are positive, we get

$$\int_0^{+\infty} e^{-2s}|v_2(s)|v_3(s)ds < +\infty. \tag{4.9}$$

Since $v_3(t)=v_3(0)e^{2t}$, (4.9) implies that

$$\int_0^{+\infty} |v_2(s)|ds < +\infty. \tag{4.10}$$

Since $z(t)<0$ by assumption, we have $v_2(s)\leq v_2(0)e^{-(N-2)s}$ for $s\geq 0$. Then for $t\geq 0$,

$$\begin{aligned}\int_t^{+\infty} e^{-2s}v_1(s)|v_2(s)|ds &= -\int_t^{+\infty} e^{-2s}v_1(s)v_2(s)ds \\ &\geq -v_2(0)\int_t^{+\infty} e^{-Ns}v_1(s)ds.\end{aligned} \tag{4.11}$$

Integrating by parts and using (2.4) we get

$$\begin{aligned}\int_t^{\infty} e^{-Ns}v_1(s)\,ds &= \frac{1}{N}e^{-Nt}v_1(t) + \frac{1}{N}\int_t^{\infty} e^{-Ns}v_1'(s)\,ds \\ &= \frac{1}{N}e^{-Nt}v_1(t) + \frac{2}{N}\int_t^{\infty} e^{-Ns}v_1(s)\,ds + \frac{1}{N}\int_t^{\infty} e^{-Ns}v_1(s)v_2(s)\,ds\end{aligned}$$

and we deduce

$$\int_t^{\infty} e^{-Ns}v_1(s) = \frac{1}{N-2}e^{-Nt}v_1(t) + \frac{1}{N-2}\int_t^{\infty} e^{-Ns}v_1(s)v_2(s)\,ds.$$

Hence for $t>0$, and since $v_2(s)<0$

$$\int_t^{\infty} e^{-Ns} v_1(s) \geq \frac{1}{N-2} e^{-Nt} v_1(t) + \frac{1}{N-2} \int_t^{\infty} e^{-2s} v_1(s) v_2(s)\, ds. \tag{4.12}$$

From (4.11) and (4.12) we have

$$\int_t^{+\infty} e^{-2s} v_1(s)|v_2(s)|ds \geq -\frac{v_2(0)}{N-2} v_1(t) e^{-Nt} + \frac{v_2(0)}{N-2} \int_t^{+\infty} v_1(s)|v_2(s)| e^{-2s} ds,$$

which implies that

$$\int_t^{+\infty} e^{-2s} v_1(s)|v_2(s)|ds \geq \frac{-v_2(0)}{N-2-v_2(0)} v_1(t) e^{-Nt}. \tag{4.13}$$

Now, from (4.6) and (4.7) we have

$$-z(t) = 2(N-2)e^{2t} \int_t^{+\infty} e^{-2s} v_1(s)|v_2(s)|ds. \tag{4.14}$$

From (4.14) and (4.13), we observe that

$$-z(t) \geq \frac{-2(N-2)v_2(0)}{N-2-v_2(0)} v_1(t) e^{(-N+2)t}. \tag{4.15}$$

Moreover, using (4.10)

$$v_1(t) = v_1(0)e^{2t}e^{\int_0^t v_2(s)ds} = v_1(0)e^{2t}e^{-\int_0^t |v_2(s)|ds} \geq v_1(0)e^{-C}e^{2t} \tag{4.16}$$

for some constant $C > 0$. Hence,

$$-z(t) \geq \frac{-2(N-2)v_1(0)v_2(0)}{N-2-v_2(0)}e^{-C}e^{(4-N)t} := C_1 e^{(4-N)t}, \tag{4.17}$$

for $C_1 > 0$, which is a contradiction with (4.6) for $N = 3, 4$.

From now on we assume $N > 4$. By the second equation in (2.4) and $z(t) = v_3(t) - 2(N-2)v_1(t)$, we get that

$$-v_2(t) = -v_2(0)e^{(2-N)t} + e^{(2-N)t}\int_0^t (-z(s))e^{(N-2)s}ds.$$

By (4.17) we have

$$\begin{aligned} |v_2(t)| &= -v_2(t) \geq -v_2(0)e^{(2-N)t} + C_1 e^{(2-N)t}\int_0^t e^{2s}ds \\ &\geq \frac{C_1}{2}e^{(2-N)t}(e^{2t}-1) \geq C_2 e^{(4-N)t}, \end{aligned}$$

for $t > 1$ where C_2 is a positive constant. Therefore,

$$\int_t^{+\infty} e^{-2s}v_1(s)|v_2(s)|ds \geq C_2\int_t^{+\infty} e^{(2-N)s}v_1(s)ds, \tag{4.18}$$

while, for N>4 and t>0

$$\int_t^{+\infty} e^{(2-N)s} v_1(s)ds = \frac{1}{N-2} v_1(t) e^{(2-N)t} - \frac{1}{N-2}\int_t^{+\infty} e^{(2-N)s} v_1(s)|v_2(s)|ds + \frac{2}{N-2}\int_t^{+\infty} e^{(2-N)s} v_1(s)ds$$
$$\geq \frac{1}{N-2} v_1(t) e^{(2-N)t} - \frac{1}{N-2}\int_t^{+\infty} e^{-2s} v_1(s)|v_2(s)|ds + \frac{2}{N-2}\int_t^{+\infty} e^{(2-N)s} v_1(s)ds.$$

So,

$$\int_t^{+\infty} e^{(2-N)s} v_1(s)ds \geq \frac{1}{N-4} v_1(t) e^{(2-N)t} - \frac{1}{N-4}\int_t^{+\infty} e^{-2s} v_1(s)|v_2(s)|ds. \tag{4.19}$$

Combining (4.18) and (4.19), we get

$$\int_t^{+\infty} e^{-2s} v_1(s)|v_2(s)|ds \geq \frac{C_2}{N-4+C_2} v_1(t) e^{(2-N)t}. \tag{4.20}$$

Then, from (4.14), (4.16) and (4.20) we obtain that

$$-z(t) \geq \frac{2(N-2)C_2 v_1(0) e^{-C}}{N-4+C_2} e^{(6-N)t} := C_3 e^{(6-N)t}, \tag{4.21}$$

for $C_3 > 0$, which is a contradiction with (4.6) for $N = 5, 6$.

Starting with (4.21) we can do the same process and obtain a contradiction for all $N \geq 3$. This ends the proof of the proposition. □

Proposition 4.2.

At any point of $W^u(P_2) \cap \{v_3 = 2(N-2)v_1\}$ *the intersection is transversal.*

Proof.

Let $V(t)=(v_1,v_2,v_3)$ be a trajectory in $W^u(P_2)$ with t in some interval $(-\infty,T)$ and $\lim_{t\to-\infty} V(t)=P_2$. Suppose that t_1 is such that $v_3(t_1)=2(N-2)v_1(t_1)$. By contradiction, assume that $V'(t_1)$ is not transversal to the plane $\{v_3(t)=2(N-2)v_1(t)\}$, that is,

$$V'(t_1) \in \{v_3 = 2(N-2)v_1\}.$$

Then, $v_3(t_1) = 2(N-2)v_1(t_1),\ v_3'(t_1) = 2(N-2)v_1'(t_1)$. From (2.4) we get $v_2(t_1)=0$. Let $z(t)=v_3(t)-2(N-2)v_1(t)$. The ODE (2.4) implies that

$$v_2' = z - (N-2)v_2, \qquad z' = 2z - 2(N-2)v_1 v_2.$$

Treating v_1 as a given function, we see that v_2,z satisfy a first order non-autonomous linear ODE and the initial condition $v_2(t_1)=0, z(t_1)=0$. Since $v_2=z=0$ is a solution of the ODE with the same initial condition, by uniqueness we deduce $v_2(t)=0$ for all t where it is defined. This contradicts $\lim_{t\to-\infty} v_2(t)=-2$. □

Proof of Theorem 1.5.

The existence of some $\lambda_*>0$ such that (1.1) has a singular solution is a consequence of Proposition 4.1. Indeed, let $V(t)=(v_1(t),v_2(t),v_3(t)):(-\infty,T)\to \mathbb{R}^3$ be the trajectory in $W^u(P_2)$ such that $v_3'(t) > 0$ as $t\to-\infty$, where T is the maximal time of existence. Then there exists some $t<T$ such that $v_3(t)\geq 2(N-2)v_1(t)$. Let t_1 be the first time such that $v_3(t_1)=2(N-2)v_1(t_1)$. Because the system (2.4) is autonomous, by shifting time, we can assume $t_1=0$. Let $P^*=V(0)$ be the point of intersection, and write $P^* = (P_1^*, P_2^*, P_3^*)$. Then

$$u(r) = -2\log(r) + \log\left(\frac{2(N-2)v_1(\log(r))}{\lambda_*}\right)$$

is a singular solution of (1.1) for $\lambda_* = P_3^*$.

The uniqueness of λ_* such that a singular solution of (1.1) exists is a consequence of Corollary 3.6, which says that singular solutions must be associated to trajectories in $W^u(P_2)$, and the trajectory in $W^u(P_2)$ with tangent vector close $(1,2,4(N-1))$ as it approaches P_2 is unique except a shift in time. This also yields the uniqueness of the singular solution.

5. MULTIPLICITY RESULT: PROOF OF THEOREM 1.6

In this section, we assume that $3 \le N \le 9$ and prove multiplicity of solutions to problem(1.1). Let $P_1=(0,0,0)$ and $P_2=(1,-2,0)$ be the stationary points of (2.4). We recall that P_1 has a 2-dimensional unstable manifold $W^u(P_1)$ and 1-dimensional stable manifold $W^s(P_1)$, while P_2 has a 1-dimensional unstable manifold $W^u(P_2)$ and a 2-dimensional stable manifold $W^s(P_2)$.

From Corollary 3.6 it follows that each regular radial solution of (1.1) corresponds to exactly one point in $W^u(P_1) \cap \{v_3=2(N-2)v_1\}$. By Proposition 4.2, we define λ_* to be the height $v_3=\lambda_*$ where $W^u(P_2)$ first intersects the plane $\{v_3=2(N-2)v_1\}$, and we denote this intersection point by

$$P^* = (P_1^*, P_2^*, P_3^*) = \left(\frac{\lambda_*}{2(N-2)}, P_2^*, \lambda_*\right). \tag{5.1}$$

Let $V_0 : R \to R^3$ be the heteroclinic connection from P_1 to P_2 contained in $\{v_3=0\}$ as stated in Proposition 2.1 and let $\hat{V_0} = V_0(-\infty, +\infty)$. Then $\hat{V_0}$ is contained in both $W^u(P_1)$ and $W^s(P_2)$.

Lemma 5.1.

$W^u(P_1)$ and $W^s(P_2)$ intersect transversally on points of $\hat{V_0}$. More precisely, for points $Q \in \hat{V_0}$ sufficiently close to P_2, there are directions in the tangent plane to $W^u(P_1)$ which are almost parallel to $v^{(1)}$, the tangent vector to $W^u(P_2)$ at P_2.

Proof.

Let u_β be the solution of the following initial value problem

$$\begin{cases} -\Delta u_\beta(r) = 2(N-2)e^{u_\beta(r)} - \beta & \text{for } 0 < r < R(\beta), \\ u_\beta(0) = 0, \qquad u'_\beta(0) = 0, \end{cases} \tag{5.2}$$

where $\beta \in \mathbb{R}$ is a parameter and $R(\beta) > 0$ is the maximal time of existence. We claim that $R(\beta) = +\infty$. Indeed, assume $R(\beta) < +\infty$ and fix $r_0 < R(\beta)$. Then for $r \in [r_0, R(\beta))$, from Eq. (5.2) we get

$$u'_\beta(r) = r_0^{N-1} u'_\beta(r_0) r^{1-N} - r^{1-N} \int_{r_0}^{r} t^{N-1} \left(2(N-2)e^{u_\beta(t)} - \beta\right) dt, \tag{5.3}$$

and this implies

$$u'_\beta(r) \leq r_0^{N-1} u'_\beta(r_0) r^{1-N} + \frac{|\beta|}{N}(r - r^{1-N} r_0^N) \quad \text{for } r_0 \leq r < R(\beta).$$

Integrating we see that

$$\limsup_{r \to R(\beta)} u_\beta(r) < +\infty.$$

Since u_β is bounded above in $[r_0, R(\beta))$, using again (5.3) we obtain

$$r_0^{N-1} u'_\beta(r_0) r^{1-N} - C(r - r^{1-N} r_0^N) \leq u'_\beta(r) \quad \text{for } r_0 \leq r < R(\beta),$$

and this shows that

$$\liminf_{r \to R(\beta)} u_\beta(r) > -\infty.$$

Control of u_β as $r \to R(\beta)$ also yields control of u'_β by (5.3) and this contradicts that $R(\beta)$ is the maximal time of existence. Therefore the solution $u_\beta(r)$ of (5.2) is defined for all $r>0$.

Let $v_\beta(t)=u_\beta(r)$ with $r=e^t$ for $t\in(-\infty,+\infty)$ and set

$$v_{1,\beta}(t) = e^{v_\beta(t)+2t}, \qquad v_{2,\beta} = v'_\beta(t), \qquad v_{3,\beta}(t) = \beta e^{2t}.$$

Then $v_{1,\beta}, v_{2,\beta}, v_{3,\beta}$ satisfies system (2.4). Let $V_\beta=(v_{1,\beta}, v_{2,\beta}, v_{3,\beta})$. We have created in this way a family of trajectories in $W^u(P_1)$ with β as a parameter. Note that for $\beta=0$, V_0 is just the heteroclinic connection of system (2.4) from P_1 to P_2 contained in the plane $\{v_3=0\}$ described in Proposition 2.1.

Define $X = \frac{\partial V}{\partial \beta}|_{\beta=0}$. Then X satisfies

$$X' = (M_2 + R(t))X$$

where M_2 is the matrix defined in (2.6) and

$$R(t) = \begin{bmatrix} v_{2,0}(t)+2 & v_{1,0}(t)-1 & 0 \\ 0 & 0 & 0 \\ 0 & 0 & 0 \end{bmatrix}.$$

Note that there exist $C,\alpha>0$, such that $|V_0(t)-P_2|\le Ce^{-\alpha t}$ for all $t\ge 0$, which follows for example from Lemma 2.2. Therefore $|R(t)|\le Ce^{-\alpha t}$ for all $t\ge 0$. Recall that the eigenvalues of M_2 are $\nu_1>0$ and ν_2,ν_3, which are complex conjugates with negative real part. Let $v^{(k)}\in C^3$ be the eigenvector associated to ν_k. By Theorem 8.1 of Chapter 3 in [25], there are solutions ψ_k to

$$\psi_k' = (M_2 + R(t))\psi_k, \quad \text{for } t > 0$$

such that $\lim_{t\to\infty}\psi_k(t)e^{-\nu_k t} = v^{(k)}$. Then

$$X(t) = \sum_{k=1}^{3} c_k \psi_k$$

for some constants $c_1, c_2, c_3 \in C$. Since ν_2, ν_3 have negative real parts, $\psi_k(t) \to 0$ as $t \to \infty$, for $k = 2,3$. If $c_1 = 0$ then $X(t) \to 0$ as $t \to \infty$ and this contradicts $\frac{\partial v_{3,\beta}}{\partial \beta}|_{\beta=0}(t) = e^{2t} > 0$ for all $t \geq 0$. So $c_1 \neq 0$ and therefore

$$X(t) = c_1 v^{(1)} e^{\nu_1 t} + o(e^{\nu_1 t}) \quad \text{as } t \to \infty.$$

This shows $X(t)$ is almost parallel to $v^{(1)}$ as $t \to \infty$. Since $v^{(1)}$ is the tangent vector to $W^u(P_2)$, then $X(t)$ is not tangent to $W^s(P_2)$ for t large. On the other hand, $X = \frac{\partial V}{\partial \beta}|_{\beta=0}$ is tangent to $W^u(P_1)$. This implies $W^s(P_2)$ and $W^u(P_1)$ intersect transversally on points of $\hat{V}_0$ close to P_2. Since the flow is invertible near $\hat{V}_0$, $W^u(P_1)$ and $W^s(P_2)$ intersect transversally at every point of $\hat{V}_0$.

We write (v_1, v_2, v_3) as points in the phase space R^3 and let $\{e_1, e_2, e_3\}$ denote the canonical basis of R^3.

We call $S \subset R^3$ a spiral around P^* if there exist independent vectors $\sigma_1, \sigma_2 \in R^3$, a continuous positive function $\rho : [0, \infty) \to R$ with $\rho(t) \to 0$ as $t \to \infty$, and $\omega \in R$ such that

$$\mathcal{S} = \{P^* + \rho(t)\cos(\omega t)\sigma_1 + \rho(t)\sin(\omega t)\sigma_2 + o(\rho(t)) : t \geq 0\}.$$

Lemma 5.2.

$W^u(P_1) \cap \{v_3 = 2(N-2)v_1\}$ *contains a spiral* S *around the point* P^*.

Proof.

The linearization of (2.4) at P_2 is given by the system

$$\begin{cases} \bar{v}_1' = \bar{v}_2, \\ \bar{v}_2' = -2(N-2)\bar{v}_1 + (2-N)\bar{v}_2 + \bar{v}_3, \\ \bar{v}_3' = 2\bar{v}_3, \end{cases}$$

which is represented by the matrix M_2. Let $\bar{M}_2$ denote the matrix

$$\bar{M}_2 = \begin{bmatrix} \mathrm{Re}(\nu_2) & -\mathrm{Im}(\nu_2) & 0 \\ \mathrm{Im}(\nu_2) & \mathrm{Re}(\nu_2) & 0 \\ 0 & 0 & \nu_1 \end{bmatrix}$$

where ν_1, ν_2 are the eigenvalues (2.7). By Lemma 2.2, system (2.4) is C^1-conjugate in a neighborhood of P_2 to the flow generated by $\bar{M}_2$ around 0. More precisely, let X_t denote the flow generated by (2.4) and $Y_t = e^{\bar{M}_2 t}$. Then there are open neighborhoods U of P_2 and V of $\bar{O} = (0,0,0)$, and a C^1 diffeomorphism $\Phi : U \to V$ such that $Y_t(x) = \Phi \circ X_t \circ \Phi^{-1}(x)$ whenever $x \in V$ and $\Phi^{-1}(x) \in U$.

Let D be the 2-dimensional disk

$$D = \left\{ V = (v_1, v_2, v_3) : v_3 = 2(N-2)v_1,\ \left| V - P^* \right| < r_0 \right\},$$

where $r_0 > 0$ is fixed and small, so that $W^u(P_2) \cap \{v_3 = 2(N-2)v_1\}$ contains only the point P^*. This $r_0 > 0$ exists by Proposition 4.2. Also by this proposition, D is transversal to $W^u(P_2)$. Let $B^s \subset W^s(P_2) \cap U \subset \{v_3 = 0\} \cap U$ be an open neighborhood of P_2 relative to $W^s(P_2)$, which is diffeomorphic to a 2-dimensional disk. Define D_t as the connected component of $X_t(D) \cap U$ that contains $X_t(P^*)$. We choose U smaller if necessary so that by the λ-Lemma of Palis [26], D_t is a C^1 manifold, which is C^1 close to B^s for t sufficiently negative. More precisely, let $\varepsilon>0$ be small to be fixed later on. Then there exists $t_0<0, |t_0|$ large, such that for all $t\leq t_0$, there is a diffeomorphism $\eta_t: D_t\to B^s$ such that $\|i'\circ\eta_t - i\|_{C^1(D_t)}\leq\varepsilon$ where i,i′ denote the inclusion maps. From now on we let $M=D_{t_0}$.

We fix $Q\in\hat{V}_0$ such that $Q\in U$ is sufficiently close to P_2. From Lemma 5.1, we can find a C^1 curve Γ contained in $W^u(P_1)$ of the form $\Gamma=\{\gamma(s): |s|<\delta_0\}$ with $\gamma:(-\delta_0,\delta_0)\to R^3$ a C^1 function such that $\gamma(0)=Q$ and $\gamma'(0)$ not tangent to $W^s(P_2)$ at Q. We can also assume that Γ is contained in U by taking δ_0 small. Choosing $\varepsilon>0$ smaller if necessary we can assume that Γ intersects M.

We want to prove that for $t>0$ large, there is a point $P_t\in X_t(\Gamma)\cap M$ and that the collection of points P_t describes a spiral around the point $X_{t_0}(P^*)$.

By the conjugation Φ, we will assume that P_2 is at the origin and near the origin the flow is given by $Y_t=e^{\tilde{M}_2 t}$. Thus the image of $W^s(P_2)\cap U$ through Φ is $\{(y_1,y_2,y_3):y_3=0\}$, which is inside V, and the image of B^s is $\{(y_1,y_2,y_3):y_3=0,|y|<\delta\}$ for some $\delta>0$.

Choosing ε small in the λ-Lemma, we can assume that the normal vector of $\tilde{\mathcal{M}}:=\Phi(\mathcal{M})$ near $\Phi(P^*)$ is almost parallel to $e_3=(0,0,1)$. Thus by taking a subset of $\tilde{\mathcal{M}}$, we may assume that $\tilde{\mathcal{M}}$ is a C^1 graph with respect to the variables (y_1,y_2), that is, there exists a C^1 function $\varphi:\{\tilde{y}=(y_1,y_2)\in\mathbb{R}^2, |\tilde{y}|<\delta\}\to\mathbb{R}$ such that

$$\tilde{\mathcal{M}}=\{(\tilde{y},\varphi(\tilde{y})):\tilde{y}\in\mathbb{R}^2, |\tilde{y}|<\delta\}.$$

Since $\gamma'(0)$ is not tangent to $W^s(P_2)$ at $\gamma(0)$, we have $\gamma_3'(0)\neq 0$. We may assume that $\varphi(\tilde{y})>0$ for $\tilde{y}$ near the origin and $\gamma_3'(0)>0$.

We claim that for all $t>0$ large there is a unique $s=s(t)>0$ small so that $Y_t(\gamma(s))\in\tilde{\mathcal{M}}$. Indeed, this condition is equivalent to

$$e^{\nu_1 t}\gamma_3(s) = \varphi(e^{\nu_2 t}(\gamma_1(s) + i\gamma_2(s))). \tag{5.5}$$

Let $\tau = 1/t > 0$ and define, for $(\tau, s) \in (0, \delta_1) \times (-\delta_1, \delta_1)$ ($\delta_1 > 0$ a small fixed number)

$$F(\tau, s) = \gamma_3(s) - e^{-\nu_1/\tau}\varphi(e^{\nu_2/\tau}(\gamma_1(s) + i\gamma_2(s))).$$

Then, since $\nu_1 > \mathrm{Re}(\nu_2)$, F admits a C^1 extension to $\tau = 0$ and

$$F(0, s) = \gamma_3(s), \qquad \frac{\partial F}{\partial \tau}(0, s) = 0, \qquad \frac{\partial F}{\partial s}(0, s) = \gamma_3'(s).$$

Since $F(0,0) = 0$ and $\frac{\partial F}{\partial s}(0,0) > 0$, by the implicit function theorem, given $t > 0$ large there is a unique s small so that $F(1/t, s) = 0$. We obtain a C^1 function $s(t) > 0$ defined for all t large such that $Y_t(\gamma(s(t))) \in \widetilde{\mathcal{M}}$. Using (5.5) we see that

$$s(t) = \frac{e^{-\nu_1 t}}{\gamma_3'(0)}\varphi(0)(1 + o(1))$$

as $t \to \infty$. Writing $\nu_2 = \alpha + i\omega$, the point of intersection has the form

$$\tilde{P}_t = Y_t(\gamma(s(t))) = (0, 0, \varphi(0, 0)) + e^{\alpha t}\cos(\omega t)\tilde{\sigma}_1 + e^{\alpha t}\sin(\omega t)\tilde{\sigma}_2 + o(e^{\alpha t})$$

where

$$\tilde{\sigma}_1 = \left(\gamma_1(0), \gamma_2(0), \frac{\partial \varphi}{\partial y_1}(0, 0)\gamma_1(0) + \frac{\partial \varphi}{\partial y_2}(0, 0)\gamma_2(0)\right)$$

$$\tilde{\sigma}_2 = \left(-\gamma_2(0), \gamma_1(0), -\frac{\partial\varphi}{\partial y_1}(0,0)\gamma_2(0) + \frac{\partial\varphi}{\partial y_2}(0,0)\gamma_1(0)\right).$$

Therefore the curve $\{\tilde{P}_t, : t > t_1\}$, where $t_1 > 0$ is large, defines a spiral contained in $\tilde{\mathcal{M}}$. Applying the conjugation Φ^{-1} we obtain a collection of points $P_t = \Phi^{-1}(\tilde{P}_t)$ in $M \cap X_t(\Gamma)$ that forms a spiral around $X_{t0}(P^*)$. Applying the flow X_{-t0} we see that

$$\mathcal{S} = \{X_{t-t_0}(\gamma(s(t))) : t \geq t_1\}$$

with $t_1 > 0$ large has the structure of a spiral around P^*. By construction S is contained in $W^u(P_1) \cap \{v_3 = 2(N-2)v_1\}$. □

Proof of Theorem 1.6.

Let us define λ_* to be the height $v_3 = \lambda_*$, where $W^u(P_2)$ first intersects the boundary plane $\{v_3 = 2(N-2)v_1\}$. Define $H_\lambda = \{v_3 = \lambda\}$. If $\lambda = \lambda_*$, we know that P^* lies on the line $\{v_3 = \lambda_*, v_3 = 2(N-2)v_1\}$. From Lemma 5.2, $W^u(P_1) \cap \{v_3 = 2(N-2)v_1\}$ contains a spiral S around the point P^*. Since the plane H_λ is transversal to $\{v_3 = 2(N-2)v_1\}$, it is possible to show that $H_{\lambda*}$ and S intersect an infinite number of times, which means that problem (1.1) has infinitely many radial regular solutions; see for example Lemma 4 in [14]. If $\lambda \neq \lambda_*$, but λ is close to λ_*, we have that $H_\lambda \cap S$ contains a large number of points, which means that problem (1.1) has a large number of radial regular solutions.

6. PROOF OF THEOREM 1.7

In this section we always assume that $N \geq 10$ and prove Theorem 1.7.

First we give the asymptotic behavior of a radial singular solution to problem (1.1) near the origin.

Lemma 6.1.

Assume that (λ_*, u_*) *is a radial singular solution of* (1.1). *Then*

$$u_*(r) = -2\log r + \log \frac{2(N-2)}{\lambda_*} + r^2 + o(r^2) \quad as\ r \to 0. \tag{6.1}$$

Proof.

By Theorem 1.4, u_* is a weakly singular radial solution of (1.1). Define $v(t)=u_*(r)$ with $r=e^t$, and v_1,v_2,v_3 are given by (2.3). Therefore, from Corollary 3.6,

$$\lim_{t\to-\infty} (v_1, v_2, v_3) = (1, -2, 0).$$

By Lemma 2.2 and Lemma 2.3, we have

$$(v_1, v_2, v_3) = (1, -2, 0) + (1, 2, 4(N-1))e^{2t}\left(1 + o(e^{\delta t})\right) \quad \text{as } t \to -\infty,$$

with $\delta > 0$ small. We then get

$$\begin{aligned} u_*(r) &= v(t) = -2t + \log \frac{2(N-2)v_1(t)}{\lambda_*} \\ &= -2\log r + \log \frac{2(N-2)\left(1 + e^{2t} + o(e^{(2+\delta)t})\right)}{\lambda_*} \\ &= -2\log r + \log \frac{2(N-2)}{\lambda_*} + \log(1 + r^2 + o(r^{2+\delta})) \\ &= -2\log r + \log \frac{2(N-2)}{\lambda_*} + r^2 + o(r^2) \quad \text{as } r \to 0. \end{aligned}$$

For $\lambda > 0$, let us define

$$w(r) = -2\log r + \log\frac{2(N-2)}{\lambda} + \frac{\lambda}{2N}r^2. \tag{6.2}$$

Let $\rho > 0$ be a small number, which will be fixed later and let us write $c_\rho = w(\rho)$. Then w satisfies

$$\begin{cases} -\Delta w \le \lambda(e^{w} - 1) & \text{in } B_\rho, \\ w(\rho) = c_\rho & \text{on } \partial B_\rho, \end{cases} \tag{6.3}$$

where B_ρ is a ball with radius ρ and center at the origin.

We have the following stability property of w.

Lemma 6.2.

Suppose $N \ge 10$ *and let* w *be defined in* (6.2). *There exists* $\rho \in (0,1)$ *small, such that* w *is stable in* B_ρ, *in the sense that*

$$\int_{B_\rho} |\nabla\varphi|^2 \ge \lambda\int_{B_\rho} e^{w}\varphi^2 \quad \textit{for all } \varphi \in C_c^\infty(B_\rho). \tag{6.4}$$

Proof.

Write $A = \frac{\lambda}{2N}$. Since $N \ge 10$,

$$\begin{aligned} \int_{B_\rho} |\nabla\varphi|^2 - \lambda e^{w}\varphi^2 &= \int_{B_\rho} |\nabla\varphi|^2 - 2(N-2)\frac{\varphi^2}{r^2}e^{Ar^2} \\ &= \int_{B_\rho}\left(|\nabla\varphi|^2 - 2(N-2)\frac{\varphi^2}{r^2}\right) - 2(N-2)(A+o(1))\int_{B_\rho}\varphi^2 \\ &\ge \int_{B_\rho}\left(|\nabla\varphi|^2 - \frac{(N-2)^2}{4}\frac{\varphi^2}{r^2}\right) - 2(N-2)(A+o(1))\int_{B_\rho}\varphi^2, \end{aligned}$$

where $o(1) \to 0$ as $\rho \to 0$. Let us recall the following improved Hardy inequality from [27]: for $\varphi \in C_c^\infty(B_\rho)$

$$\int_{B_\rho}\left(|\nabla\varphi|^2-\frac{(N-2)^2}{4}\frac{\varphi^2}{r^2}\right)\geq H_2\rho^{-2}\int_{B_\rho}\varphi^2,$$

where the constant H_2 is the first eigenvalue of the Laplacian in the unit ball in $N=2$, hence it is positive and independent of N.

Choose $\rho>0$ such that $2(N-2)(A+o(1))\leq H_2\rho^{-2}$. Then (6.4) holds. □

Lemma 6.3.

Let $\rho\in(0,1)$ *be small and satisfy Lemma* 6.2. *Then for any radial regular solution* u *of* (1.1) *we have*

$$u(r)\leq\begin{cases}w(r) & in\ B_\rho\\ c_\rho & in\ B\setminus B_\rho,\end{cases}\tag{6.5}$$

where $w(r)$ *is defined in* (6.2).

Proof.

Arguing by contradiction, suppose there exists $r_0\in(0,\rho)$, such that $u(r_0)=w(r_0)$. Then

$$\begin{cases}-\Delta u=\lambda(e^u-1) & \text{in } B_{r_0};\\ -\Delta w\leq\lambda(e^w-1) & \text{in } B_{r_0};\\ u=w & \text{on } \partial B_{r_0}.\end{cases}\tag{6.6}$$

Therefore,

$$\begin{cases}-\Delta(w-u)\leq\lambda\left(e^w-e^u\right) & \text{in } B_{r_0},\\ w-u=0 & \text{on } \partial B_{r_0}.\end{cases}\tag{6.7}$$

Multiplying by $(w-u)^+$ and integrating in (6.7), we obtain

$$\int_{B_{r_0}} |\nabla(w-u)^+|^2 \leq \lambda \int_{B_{r_0}} (e^w - e^u)(w-u)^+. \tag{6.8}$$

From Lemma 6.2, w is stable in B_{r0}, by taking $\varphi = (w-u)^+$ in (6.4), we then have

$$\int_{B_{r_0}} |\nabla(w-u)^+|^2 - \lambda e^w((w-u)^+)^2 \geq 0. \tag{6.9}$$

Combining (6.8) and (6.9), we get

$$\lambda \int_{B_{r_0}} e^w((w-u)^+)^2 \leq \lambda \int_{B_{r_0}} (e^w - e^u)(w-u)^+.$$

We rewrite it as

$$\int_{B_{r_0}} \left[(e^w - e^u)(w-u)^+ - e^w((w-u)^+)^2\right] \geq 0.$$

By convexity, the integrand is nonpositive, therefore,

$$(e^w - e^u)(w-u)^+ - e^w((w-u)^+)^2 = 0 \quad \text{a.e. in } B_{r_0},$$

then

$$(w-u)^+ = 0 \quad \text{a.e. in } B_{r_0}.$$

It implies that $w \leq u$ in B_{r0}, which is impossible because u is a radial regular solution. Then $u(r) \leq w(r)$ for $r \in (0,\rho)$.

Since u is a radially decreasing regular solution, $u \leq c_\rho$ in $B \setminus B_\rho$. □

Now, let (λ, u_λ) be any radial solution to (1.1) (regular or singular), and define the operator L_γ as

$$L_\gamma(\phi) = -\Delta\phi - \lambda e^{u_\lambda}\phi + \gamma\phi$$

with $\gamma > 0$ large but fixed. We have the following lemma.

Lemma 6.4.

If $\gamma > 0$ is fixed large enough, we have:

(a)*for* $N \geq 11$, $\langle L_\gamma(\phi), \phi\rangle \geq C_1\|\phi\|^2_{H_0^1(B)}$ *for all* $\phi \in C_c^\infty(B)$;

(b)*for* $N = 10$, $\langle L_\gamma(\phi), \phi\rangle \geq C_2\|\phi\|^2_{L^2(B)}$ *for all* $\phi \in C_c^\infty(B)$,

where C_1 *and* C_2 *are positive constants.*

Proof.

For $\rho > 0$ small given in Lemma 6.2, from Lemma 6.1 and Lemma 6.3, we have

$$\begin{aligned}
\langle L_\gamma(\phi), \phi\rangle &= \int_B L_\gamma(\phi)\phi = \int_B \left(|\nabla\phi|^2 - \lambda e^{u_\lambda}\phi^2 + \gamma\phi^2\right) \\
&= \int_B |\nabla\phi|^2 - \int_{B_\rho} \lambda e^{u_\lambda}\phi^2 - \int_{B\setminus B_\rho} \lambda e^{u_\lambda}\phi^2 + \int_B \gamma\phi^2 \\
&\geq \int_B |\nabla\phi|^2 - 2(N-2)\int_{B_\rho} \frac{\phi^2}{r^2}(1 + Ar^2 + o(r^2)) - C\int_{B\setminus B_\rho} \phi^2 + \int_B \gamma\phi^2 \\
&\geq \int_B \left(|\nabla\phi|^2 - 2(N-2)\frac{\phi^2}{r^2}\right) + [\gamma - \max\{2(N-2)(A + o(1)), C\}]\int_B \phi^2
\end{aligned}$$

where $A = \frac{\lambda}{2N}$ for a radial regular solution u_λ, $A = 1$ for a radial singular solution u_λ, and $o(1) \to 0$ as $\rho \to 0$. Choose γ large such that the second term of above is nonnegative, we then get the conclusion by Hardy's inequality. □

We now define

$$\|\phi\|_H^2 := \int_B \left(|\nabla\phi|^2 - \lambda e^{u_\lambda}\phi^2 + \gamma\phi^2\right)$$

which is a norm on $C_c^\infty(B)$ with associated inner product

$$(\phi,\varphi)_H = \int_B \left(\nabla\phi\nabla\varphi - \lambda e^{u_\lambda}\phi\varphi + \gamma\phi\varphi\right).$$

Completing $C_c^\infty(B)$ with respect to this norm we obtain a Hilbert space H. We denote by H^* the dual of H. We have $H_0^1(B) \subset H \subset L^2(B)$ and therefore $L^2(B) \subset H^* \subset H^{-1}(B)$. Actually by Lemma 6.4, if $N \geq 11$, the space H is just $H_0^1(B)$.

Given $h \in L^2(B) \subset H^*$ we consider the following problem

$$L_\gamma\phi = h \quad \text{in } B, \quad \text{and} \quad \phi = 0 \quad \text{on } \partial B. \tag{6.10}$$

We say that $\phi \in H$ is a weak solution of problem (6.10) if

$$(\phi,\varphi)_H = \langle h,\varphi\rangle_{H^*,H} \quad \text{for all } \varphi \in H.$$

By the Lax–Milgram theorem, for $h \in L^2(B)$, problem (6.10) has a unique weak solution $\phi \in H$.

Lemma 6.5.

Let $T : L^2(B) \to L^2(B)$ *be the operator defined by* $Th = \phi$, *where* ϕ *is the solution of* (6.10). *Then* T *is compact and the natural embedding* $H \hookrightarrow L^2(B)$ *is compact.*

Proof.

For $N \geq 11$, both statements hold since $T : L^2(B) \to H = H_0^1(B)$ and $H_0^1(B) \hookrightarrow L^2(B)$ is compact, by the Rellich–Kondrachov theorem. For $N = 10$, we observe that L_γ satisfies

$$\langle L_\gamma(\phi), \phi \rangle \geq c_r \|\phi\|_{L^r(B)}^2 \quad \forall \phi \in C_c^\infty(B)$$

for $2 \leq r < \frac{2N}{N-2}$ where $c_r > 0$, thanks to an improved Hardy inequality of Brezis and Vázquez [27]. Then the statements are proved in [28]. □

Proposition 6.6.

The radial singular solution (λ_*, u_*) *of* (1.1) *has a finite Morse index.*

Proof.

By Lemma 6.5, if $\gamma > 0$ is large, $(-\Delta - \lambda_* e^{u_*} + \gamma)^{-1}$ is well defined and compact from $L^2(B)$ into itself, and hence its spectrum except 0 consists of eigenvalues, and these eigenvalues form a sequence that converges to 0. Hence $-\Delta - \lambda_* e^{u_*}$ is negative definite on a finite dimensional space only. □

Next we prove a bound for the Morse index of any radial regular solution of (1.1).

Proposition 6.7.

There is an integer $K \geq 1$ *independent of* λ, *such that for any radial regular solution* u_λ *of* (1.1) *we have*

$$1 \leq m(u_\lambda) \leq K, \tag{6.11}$$

where $m(u_\lambda)$ *denotes the Morse index of* u_λ.

Proof.

From (1.1) we get

$$\int_B |\nabla u_\lambda|^2 = \lambda \int_B (e^{u_\lambda} - 1)u_\lambda .$$

Therefore,

$$\int_B \left(|\nabla u_\lambda|^2 - \lambda e^{u_\lambda} u_\lambda^2\right) = \lambda \int_B \left(e^{u_\lambda} - 1 - e^{u_\lambda} u_\lambda\right) u_\lambda < 0,$$

so $m(u_\lambda) \geq 1$.

We prove the proposition by contradiction. Suppose that $\{(\lambda_n, u_n)\}$ is a sequence of radial regular solutions of problem (1.1) and assume that $m(u_n) \to \infty$ as $n \to \infty$. Let us write $m(u_n) = m_n$ and

$$L_n = -\Delta - \lambda_n e^{u_n} .$$

Let

$$E_n = \text{span}\left\{\varphi \in L^2(B) : \varphi \text{ is eigenvector of } L_n \text{ with negative eigenvalue}\right\}$$

so that $\dim(E_n) = m_n$. Since L_n is symmetric there exist eigenfunctions $\varphi_{1,n}, \ldots, \varphi_{m_n,n} \in E_n$, namely

$$\begin{cases} L_n \varphi_{i,n} = \mu_{i,n} \varphi_{i,n} & \text{in } B, \\ \varphi_{i,n} = 0 & \text{on } \partial B, \end{cases}$$

with $\mu_{i,n} < 0$, that form an orthonormal basis of E_n in $L^2(B)$ sense, that is

$$\int_B \varphi_{i,n}\varphi_{j,n} = \delta_{ij} \quad \text{for } i, j \in \{1, 2, \ldots, m_n\}, \tag{6.12}$$

where δ_{ij} is Kronecker's delta.

Multiplying by $\varphi_{i,n}$ and integrating on B, we find

$$\int_B \left(|\nabla\varphi_{i,n}|^2 - \lambda_n e^{u_n}\varphi_{i,n}^2\right) = \mu_{i,n}\int_B \varphi_{i,n}^2 < 0.$$

Then

$$\begin{aligned}
\int_B |\nabla\varphi_{i,n}|^2 &< \int_B \lambda_n e^{u_n}\varphi_{i,n}^2 = \int_{B_\rho} \lambda_n e^{u_n}\varphi_{i,n}^2 + \int_{B\setminus B_\rho} \lambda_n e^{u_n}\varphi_{i,n}^2 \\
&\le \int_{B_\rho} \lambda_n e^{-2\log r + \log\frac{2(N-2)}{\lambda_n} + A_n r^2}\varphi_{i,n}^2 + C\int_{B\setminus B_\rho} \varphi_{i,n}^2 \\
&= 2(N-2)\int_{B_\rho} \frac{\varphi_{i,n}^2}{r^2}(1 + A_n r^2 + o(r^2)) + C\int_{B\setminus B_\rho} \varphi_{i,n}^2 \\
&\le \frac{8}{N-2}\int_B |\nabla\varphi_{i,n}|^2 + \max\{2(N-2)(A_n + o(1)), C\}\int_B \varphi_{i,n}^2.
\end{aligned}$$

If $N \ge 11$ we deduce

$$\int_B |\nabla\varphi_{i,n}|^2 \le \frac{N-2}{N-10}\max\{2(N-2)(A_n + o(1)), C\},$$

where $A_n = \frac{\lambda_n}{2N}$. Let us assume $N \ge 11$ and leave the case $N = 10$ for later. Thus $(\varphi_{i,n})_n$ is bounded in $H_0^1(B)$. By a diagonal argument, there is a subsequence (which we write the same), such that for each $i \in \{1, 2, \ldots\}$, $\varphi_{i,n} \rightharpoonup \varphi_i$ weakly in $H_0^1(B)$, $\varphi_{i,n} \to \varphi_i$ strongly in $L^2(B)$ and almost everywhere in B as $n \to +\infty$. Therefore for all $i \ge 1$,

$$\|\varphi_i\|_{H_0^1(B)} \leq \liminf_{n\to+\infty} \|\varphi_{i,n}\|_{H_0^1(B)} \leq C, \qquad \|\varphi_i\|_{L^2(B)} = 1.$$

Moreover, taking $n \to \infty$ in (6.12)

$$\int_B \varphi_i \varphi_j = \delta_{ij} \quad \text{for } i, j \geq 1. \tag{6.13}$$

Since $(\varphi_i)_{i\geq 1}$ is bounded in $H_0^1(B)$, there is a subsequence $(\varphi_{i_j})_j$ of (φ_i) such that $\varphi_{i_j} \to \varphi$ in $L^2(B)$ as $j \to +\infty$, and $\|\varphi\|_{L^2(B)} = 1$. But from (6.13) we get

$$\int_B \varphi_{i_j} \varphi_{i_m} = 0 \quad \text{for } j \neq m.$$

Taking the limit, as $j \to +\infty$ and $m \to +\infty$, we have

$$\int_B \varphi^2 = 0,$$

which is a contradiction.

For $N=10$, we define the Hilbert space H as the completion of $C_c^\infty(B)$ with respect to the norm

$$\|\phi\|_H^2 := \int_B \left(|\nabla\phi|^2 - \lambda_* e^{u_*}\phi^2 + \gamma\phi^2 \right)$$

with $\gamma > 0$ large but fixed and u_* the radial singular solution of (1.1) with $\lambda = \lambda_*$. Then

$$
\begin{aligned}
\|\varphi_{i,n}\|_H^2 &= \int_B \left(|\nabla \varphi_{i,n}|^2 - \lambda_* e^{u_*} \varphi_{i,n}^2\right) + \gamma \int_B \varphi_{i,n}^2 \\
&= \mu_{i,n} \int_B \varphi_{i,n}^2 + \int_B \left(\lambda_n e^{u_n} - \lambda_* e^{u_*}\right) \varphi_{i,n}^2 + \gamma \int_B \varphi_{i,n}^2 \\
&< \int_B \left(\lambda_n e^{u_n} - \lambda_* e^{u_*}\right) \varphi_{i,n}^2 + \gamma \int_B \varphi_{i,n}^2 .
\end{aligned}
$$

Let $\rho > 0$ be as in Lemma 6.2. Let $A_n = \frac{\lambda_n}{2N}$. From Lemma 6.1 and Lemma 6.3, we find

$$
\begin{aligned}
\int_B \left(\lambda_n e^{u_n} - \lambda_* e^{u_*}\right) \varphi_{i,n}^2 &= \int_{B_\rho} \left(\lambda_n e^{u_n} - \lambda_* e^{u_*}\right) \varphi_{i,n}^2 + \int_{B \setminus B_\rho} \left(\lambda_n e^{u_n} - \lambda_* e^{u_*}\right) \varphi_{i,n}^2 \\
&\le \int_{B_\rho} \left(\lambda_n e^{-2\log r + \log \frac{2(N-2)}{\lambda_n} + A_n r^2} - \lambda_* e^{-2\log r + \log \frac{2(N-2)}{\lambda_*} + r^2 + o(r^2)}\right) \varphi_{i,n}^2 + C \int_{B \setminus B_\rho} \varphi_{i,n}^2 \\
&\le C \int_B \varphi_{i,n}^2 .
\end{aligned}
$$

Thus we get

$$
\|\varphi_{i,n}\|_H^2 \le (C + \gamma) \int_B \varphi_{i,n}^2 \le C.
$$

That is, $(\varphi_{i,n})_n$ is bounded in H. By Lemma 6.5, the natural embedding $H \hookrightarrow L^2(B)$ is compact, so using the same argument as the case $N \ge 11$ we obtain a contradiction. This ends the proof of Proposition 6.7. □

Lemma 6.8.

Suppose that u_1, u_2 *are radial regular solutions of* (1.1) *associated to the same parameter* $\lambda > 0$. *Then the graph of* u_1 *must intersect with the graph of* u_2.

Proof.

By contradiction, assume that $u_1(r) > u_2(r)$ for any $r \in (0,1)$, and set $v = u_1 - u_2$. By Eq. (1.1) we have

$$\begin{cases} -\Delta v = \lambda(e^{u_1} - e^{u_2}) > \lambda e^{u_2} v & \text{in } B; \\ v > 0 & \text{in } B; \\ v = 0 & \text{on } \partial B. \end{cases} \tag{6.14}$$

We consider the following eigenvalue problem

$$\begin{cases} -\Delta \psi = \lambda e^{u_2} \psi + \mu \psi & \text{in } B; \\ \psi > 0 & \text{in } B; \\ \psi = 0 & \text{on } \partial B. \end{cases} \tag{6.15}$$

Multiplying by ψ and v in (6.14) and (6.15) respectively, and then integrating on B, we get

$$\lambda \int_B e^{u_2} \psi v + \mu \int_B \psi v > \lambda \int_B e^{u_2} \psi v,$$

so $\mu > 0$, that is u_2 is a stable radial regular solution. Then $m(u_2) = 0$ and this contradicts Proposition 6.7. □

Proof of Theorem 1.7.

The first part follows from Proposition 6.6 and Proposition 6.7.

Let K be an integer such that $m(u_\lambda) \leq K$ for any radial regular solution u_λ of (1.1) and $m(u_*) \leq K$. This integer exists by Proposition 6.6 and Proposition 6.7. Next we prove that the graph of any radial regular solution u_λ of (1.1) intersects with that of the radial singular solution u_* at most $2K+1$ times in $(0,1)$. We follow the idea of Theorem 1.2 in [10].

By contradiction, suppose that the graph of u_λ intersects with the graph of u_* at least $2K+2$ times in $(0,1)$. There are two cases: $\lambda < \lambda_*$ and $\lambda \geq \lambda_*$.

For $\lambda < \lambda_*$, we can show $m(u_\lambda) \geq K+1$, contradicting Proposition 6.7. Indeed, since the graph of (λ, u_λ) intersects with that of (λ_*, u_*) at least $2K+2$ times in $(0,1)$, there are at least $K+1$ intervals $J_i \subset (0,1)$ $(i=1,2,\ldots,K+1)$ such that $u_\lambda > u_*$ in J_i. Let

$$h_i = \begin{cases} u_\lambda - u_* & \text{in } J_i; \\ 0 & \text{in } (0,1)\backslash J_i. \end{cases}$$

Since u_λ and u_* satisfy Eq. (1.1), we have

$$\begin{aligned} -\Delta(u_\lambda - u_*) &= \lambda(e^{u_\lambda} - 1) - \lambda_*(e^{u_*} - 1) \\ &< \lambda(e^{u_\lambda} - e^{u_*}) \le \lambda e^{u_\lambda}(u_\lambda - u_*). \end{aligned}$$

Therefore

$$Q_{u_\lambda}(h_i) = \int_B [|\nabla h_i|^2 - \lambda e^{u_\lambda} h_i^2] dx < 0.$$

Since the functions h_i, $i = 1, \ldots, K+1$ are linearly independent, we conclude that $m(u_\lambda) \ge K+1$.

For $\lambda \ge \lambda_*$, similarly we can obtain that $m(u_*) \ge K+1$. This contradicts Proposition 6.6. In fact, because the graph of u_λ intersects with that of u_* at least $2K+2$ times in $(0,1)$, there are at least $K+1$ intervals $J_k \subset (0,1)$ $(k = 1, 2, \ldots, K+1)$ such that $u_* > u_\lambda$ in J_k. Let

$$h_k = \begin{cases} u_* - u_\lambda & \text{in } J_k; \\ 0 & \text{in } (0,1)\backslash J_k. \end{cases}$$

Note that

$$-\Delta h_k < \lambda_* e^{u_*} h_k \quad \text{in } J_k$$

and this implies

$$Q_{u_*}(h_k) = \int_B [|\nabla h_k|^2 - \lambda_* e^{u_*} h_k^2] dx < 0.$$

Therefore $m(u_*) \geq K+1$.

Next we prove that the number of regular solutions to (1.1) is bounded by $(K+1)^2$ for each $\lambda \in (\lambda_0, \mu_1)$.

By contradiction, for each fixed $\lambda \in (\lambda_0, \mu_1)$, we suppose that there are at least $(K+1)^2+1$ radial regular solutions to (1.1), denoted by u_i $(i = 0, 1, \ldots, (K+1)^2)$. Without loss of generality, assume $u_0(0) > u_1(0) > \cdots > u_{(K+1)^2}(0)$. By Lemma 6.8, the graph of u_i, $i = 1, \ldots, (K+1)^2$, must intersect with that of u_0. Let a_i be the first point such that $u_i(a_i) = u_0(a_i)$ for $i = 1, \ldots, (K+1)^2$. Then there are the following two cases.

Case 1: There are at least $(K+1)$ different points a_i such that $u_0 - u_i > 0$ in $(0, a_i)$ and $u_i(a_i) = u_0(a_i)$.

Case 2: There exists some point $a_{i0} \in (0,1)$, such that there are at least $(K+1)$ regular solutions that first intersect u_0 at a_{i0}.

Case 1. We rearrange the indices so that $a_1 < \cdots < a_{K+1}$. Now $u_1(0), \ldots, u_{K+1}(0)$ are not necessarily ordered. Let $\varphi_i = (u_0 - u_i)\chi_{(0, a_i)}$. We claim that $\{\varphi_i : i = 1, 2, \ldots, (K+1)\}$ is linearly independent. Indeed, suppose that

$$\sum_{i=1}^{K+1} c_i \varphi_i = 0.$$

Since $a_{i-1} < a_i$, there exists $r_{i-1} \in (a_{i-1}, a_i)$, such that $\varphi_1(r_{i-1}) = 0$, $\varphi_2(r_{i-1}) = 0, \ldots, \varphi_{i-1}(r_{i-1}) = 0, \varphi_i(r_{i-1}) \neq 0$, then we can get $c_i = 0$, for $i = 1, 2, \ldots, (K+1)$. Then

$$Q_{u_0}(\varphi_i) = \int_{\{|x|<a_i\}} [|\nabla\varphi_i|^2 - \lambda e^{u_0}\varphi_i^2]dx$$
$$= \lambda \int_{\{|x|<a_i\}} [e^{u_0} - e^{u_i} - e^{u_0}(u_0 - u_i)](u_0 - u_i)dx < 0$$

by strict convexity and $u_0 - u_i > 0$ in $\{|x| < a_i\}$. This implies that $m(u_0) \geq K+1$, contradicting Proposition 6.7.

Case 2. Rearranging indices, there are at least $K+1$ solutions $u_1, \ldots, u_{K+1}$ that satisfy$(u_0(r) - u_j(r)) > 0$ for $r \in (0, a_{i0})$ and $u_j(a_{i0}) = u_0(a_{i0}), j = 1, \ldots, K+1$. Set $\varphi_j = (u_0 - u_j)\chi_{(0,a_{i0})}$, we claim that

$$\{\varphi_j : j = 1, \ldots, K+1\} \text{ is linearly independent.} \tag{6.16}$$

Claim (6.16) together with $Q_{u_0}(\varphi_j) < 0$ yields that $m(u_0) \geq K+1$, contradicting$1 \leq m(u_0) \leq K$.

Let us show that the claim (6.16) holds. From now on, we write $r_0 = a_{i0}$. We assume that there exist c_j, $j = 1, \ldots, K+1$, such that

$$\sum_{j=1}^{K+1} c_j\varphi_j(r) = 0 \quad \text{for all } r \in (0, r_0],$$

that is,

$$\sum_{j=1}^{K+1} c_j u_j(r) = \left(\sum_{j=1}^{K+1} c_j\right) u_0(r) \quad \text{for all } r \in (0, r_0]. \tag{6.17}$$

We will deduce $c_1 = \cdots = c_{K+1} = 0$ from the following assertion:

$$\sum_{j=1}^{K+1} c_j (u_j'(r_0))^n = \left(\sum_{j=1}^{K+1} c_j\right)(u_0'(r_0))^n, \quad \text{for all integers } n \geq 0. \tag{6.18}$$

In the following we will establish (6.18). We denote $g^{(n)}$ the n-th derivative of g and set

$$f(u) := -\lambda(e^u - 1), \quad \forall u \in \mathbb{R}; \qquad b = u_0(r_0).$$

Then $f^{(n)}(u_j(r_0)) = -\lambda e^b$ for any integer $n \geq 1$.

In order to prove (6.18), we shall show that for each $j \in \{0, 1, 2, \ldots, K+1\}$,

$$u_j^{(n)}(r_0) = P_n(u'_j(r_0)) \quad \text{for any integer } n \geq 1, \tag{6.19}$$

where P_n is a polynomial of degree 1 for $n = 1, 2$, and of degree $n-2$ for $n \geq 3$, whose coefficients depend only on N, n, r_0, and b.

Indeed, for $n = 1$, (6.19) is direct and for $n = 2$ this follows from Eq. (1.1). By induction, assume that (6.19) holds for $n = k \geq 2$. From Eq. (1.1), we have

$$(\Delta u_j)^{(k-1)} = (f(u_j))^{(k-1)}. \tag{6.20}$$

We see that for $n \geq 0$,

$$\begin{aligned}(\Delta u_j)^{(n)} &= u_j^{(n+2)} + \frac{N-1}{r} u_j^{(n+1)} - n\frac{N-1}{r^2} u_j^{(n)} + n(n-1)\frac{N-1}{r^3} u_j^{(n-1)} - \cdots \\ &\quad + (-1)^{n-1} n! \frac{N-1}{r^n} u_j'' + (-1)^n n! \frac{N-1}{r^{n+1}} u_j',\end{aligned} \tag{6.21}$$

and by the formula for derivatives of a composition (e.g. Faa di Bruno [29]) we obtain

$$(f(u_j))^{(n)} = -\lambda e^{u_j} \sum_{\alpha_1,\ldots,\alpha_n} \frac{n!}{\alpha_1!(1!)^{\alpha_1}\alpha_2!(2!)^{\alpha_2}\cdots\alpha_n!(n!)^{\alpha_n}} \prod_{i=1}^{n} (u_j^{(i)})^{\alpha_i}, \tag{6.22}$$

where the sum ranges over integers $\alpha_1 \geq 0$, with $\alpha_1 + 2\alpha_2 + \cdots + n\alpha_n = n$. Using (6.20), (6.21) and (6.22) with $n = k-1$ and $r = r_0$, we get

$$\begin{aligned} u_j^{(k+1)}(r_0) &= -\frac{N-1}{r_0} u_j^{(k)}(r_0) + (k-1)\frac{N-1}{r_0^2} u_j^{(k-1)}(r_0) - \cdots \\ &\quad - (-1)^{k-2}(k-1)!\frac{N-1}{r_0^{k-1}} u_j''(r_0) - (-1)^{k-1}(k-1)!\frac{N-1}{r_0^k} u_j'(r_0) \\ &\quad - \lambda e^b \sum_{\alpha_1,\ldots,\alpha_{k-1}} \frac{(k-1)!}{\alpha_1!(1!)^{\alpha_1}\alpha_2!(2!)^{\alpha_2}\cdots\alpha_{k-1}!((k-1)!)^{\alpha_{k-1}}} \prod_{i=1}^{k-1} (u_j^{(i)}(r_0))^{\alpha_i}, \end{aligned}$$

where the sum ranges over integers $\alpha_1 \geq 0, \ldots, \alpha_{k-1} \geq 0$ with $\alpha_1 + 2\alpha_2 + \cdots + (k-1)\alpha_{k-1} = k-1$. By the induction assumption (6.19), we have $\prod_{i=1}^{k-1} (u_j^{(i)}(r_0))^{\alpha_i}$ is a polynomial in $u_j'(r_0)$ of degree at most $\alpha_1 + \alpha_2 + \alpha_3 + 2\alpha_4 + 3\alpha_5 + \cdots + (k-3)\alpha_{k-1} \leq k-1$. Thus we see the validity of (6.19).

Next we prove that (6.18) holds, again by induction. From (6.17), we have

$$\sum_{j=1}^{K+1} c_j u_j^{(n)}(r_0) = \left(\sum_{j=1}^{K+1} c_j\right) u_0^{(n)}(r_0) \quad \text{for any integer } n \geq 0, \tag{6.23}$$

and so (6.18) holds for $n = 0, 1$. Suppose (6.18) holds for $n = k$. By Eq. (1.1), we get

$$(\Delta u_j)^{(n)} = (f(u_j))^{(n)}. \tag{6.24}$$

Since $u_j(r_0)=u_0(r_0)$ for $j=1,2,\ldots,K+1$,
from (6.21), (6.22), (6.23) and (6.24), we obtain for any integer $n\geq 0$,

$$\sum_{j=1}^{K+1} c_j\left(\left(u_j'(r_0)\right)^n + A_{j,n}\right) = \left(\sum_{j=1}^{K+1} c_j\right)\left((u_0'(r_0))^n + A_{0,n}\right) \tag{6.25}$$

where

$$A_{j,n} = \sum_{\alpha_1,\ldots,\alpha_n} \frac{n!}{\alpha_1!(1!)^{\alpha_1}\alpha_2!(2!)^{\alpha_2}\cdots\alpha_n!(n!)^{\alpha_n}} \prod_{i=1}^{n}(u_j^{(i)}(r_0))^{\alpha_i}$$

and the sum ranges over integers $0\leq\alpha_1<n,\alpha_2\geq 0,\ldots,\alpha_n\geq 0$ with $\alpha_1+2\alpha_2+\cdots+n\alpha_n=n$. In writing (6.25) we have used again the formula for the n-th order derivative of a composition, where we have isolated one term. Consider (6.25) for $n=k+1$. By (6.19) we know that $\prod_{i=1}^{k+1}(u_j^{(i)}(r_0))^{\alpha_i}$ is a polynomial in $u_j'(r_0)$ of degree at most

$$\alpha_1+\alpha_2+\alpha_3+2\alpha_4+3\alpha_5+\cdots+(k-1)\alpha_{k+1}.$$

Since $0\leq\alpha_1<k+1$, we see that

$$\alpha_1+\alpha_2+\alpha_3+2\alpha_4+3\alpha_5+\cdots+(k-1)\alpha_{k+1}<\alpha_1+2\alpha_2+\cdots+(k+1)\alpha_{k+1}=k+1$$

and therefore $A_{j,n}$ can be expressed as a polynomial in $u_j'(r_0)$ of degree at most k. Thus by the induction assumption, we have

$$\sum_{j=1}^{K+1} c_j A_{j,n} = \left(\sum_{j=1}^{K+1} c_j\right) A_{0,n}$$

and so (6.18) holds for any integer $n \geq 0$.

Finally we turn to the proof of (6.16), namely the linear independence of $\varphi_j, j=1,\ldots,K+1$. We denote $u_0'(r_0) = d_0, u_j'(r_0) = d_j$ for $j=1,2,\ldots,K+1$. For $n=1,2,\ldots,K+1$, we can rewrite (6.18) as

$$\begin{pmatrix} d_1 - d_0 & d_2 - d_0 & \cdots & d_{K+1} - d_0 \\ d_1^2 - d_0^2 & d_2^2 - d_0^2 & \cdots & d_{K+1}^2 - d_0^2 \\ d_1^3 - d_0^3 & d_2^3 - d_0^3 & \cdots & d_{K+1}^3 - d_0^3 \\ \vdots & \vdots & \ddots & \vdots \\ d_1^{K+1} - d_0^{K+1} & d_2^{K+1} - d_0^{K+1} & \cdots & d_{K+1}^{K+1} - d_0^{K+1} \end{pmatrix} \begin{pmatrix} c_1 \\ c_2 \\ c_3 \\ \vdots \\ c_{K+1} \end{pmatrix} = 0. \tag{6.26}$$

A calculation shows that the determinant of the coefficient matrix of (6.26) is equal to a $(K+2) \times (K+2)$ Vandermonde determinant and the value is

$$\prod_{0 \leq j < i \leq K+1} (d_i - d_j) \neq 0.$$

Thus $c_1 = c_2 = \cdots = c_{K+1} = 0$ and this ends the proof of Theorem 1.7. □

ACKNOWLEDGMENTS

J. Dávila was supported by Fondecyt1090167, CAPDE-Anillo ACT-125 and Fondo Basal CMM. W. Chen was supported by a doctoral grant of CONICYT Chile. The authors thank the referee for pointing out Ref. [6].

APPENDIX.

Proof of Proposition 1.2.

Suppose u is a classical solution of (1.1). Let $\phi_1 > 0$ be the first eigenfunction of $-\Delta$ corresponding to the first eigenvalue μ_1. Multiplying problem (1.1) by ϕ_1 and integrating over B, we find

$$\mu_1 \int_B u\phi_1 = \lambda \int_B (e^u - 1)\phi_1 > \lambda \int_B u\phi_1.$$

Thus $\lambda < \mu_1$.

Multiplying problem (1.1) by $x \cdot \nabla u$, and integrating over B, we have

$$-\int_B \Delta u(x \cdot \nabla u) = \lambda \int_B (e^u - 1)(x \cdot \nabla u). \tag{A.1}$$

But

$$\begin{aligned} -\int_B \Delta u(x \cdot \nabla u) &= -\frac{1}{2}\int_{\partial B} |\nabla u|^2 x \cdot \nu + \left(1 - \frac{N}{2}\right)\int_B |\nabla u|^2 \\ &\leq \left(1 - \frac{N}{2}\right)\int_B |\nabla u|^2, \end{aligned} \tag{A.2}$$

since $x \cdot \nu \geq 0$ on ∂B. Moreover,

$$\lambda \int_B (e^u - 1)(x \cdot \nabla u) = -\lambda N \int_B (e^u - 1 - u).$$

From (A.1), (A.2) and (A.3), we get

$$\left(\frac{N}{2}-1\right)\int_B |\nabla u|^2 \le \lambda N \int_B (e^u - 1 - u).$$

We rewrite the above inequality as

$$\frac{N-2}{4}\int_B |\nabla u|^2 \le \lambda N \int_B (e^u - 1 - u) - \frac{N-2}{4}\int_B |\nabla u|^2.$$

Multiplying Eq. (1.1) by u and substituting we get

$$\frac{N-2}{4}\int_B |\nabla u|^2 \le \lambda \int_B \left[N(e^u - 1 - u) - \frac{N-2}{4}(e^u - 1)u\right].$$

The integrand on the right hand is negative for $u \ge C_0$, with C_0, a positive constant, so the integral can be restricted to the region $\{x : u(x) \le C_0\}$ and in this region

$$N(e^u - 1 - u) - \frac{N-2}{4}(e^u - 1)u \le C_1 u^2.$$

Thus

$$\frac{N-2}{4}\int_B |\nabla u|^2 \le \lambda C_1 \int_B u^2 \le \lambda C_2 \int_B |\nabla u|^2,$$

where $C_1 > 0, C_2 > 0$. This implies that $u = 0$ if $0 < \lambda < \frac{N-2}{4C_2}$. □

REFERENCES

1. B. Gidas, W. Ni, L. Nirenberg, Symmetry and related properties via the maximum principle, Comm. Math. Phys. 68 (3) (1979) 209–243.
2. I.M. Gelfand, Some problems in the theory of quasilinear equations, Amer. Math. Soc. Transl. 29 (2) (1963) 295–381.
3. D.D. Joseph, T.S. Lundgren, Quasilinear Dirichlet problems driven by positive sources, Arch. Ration. Mech. Anal. 49 (1972–1973) 241–269.
4. K. Nagasaki, T. Suzuki, Spectral and related properties about the Emden–Fowler equation −1u = λeu on circular domains, Math. Ann. 299 (1) (1994)
5. 1–15.
6. L. Dupaigne, Stable Solutions of Elliptic Partial Differential Equations, in: Chapman & Hall/CRC Monographs and Surveys in Pure and Applied Mathematics, vol. 143, Chapman & Hall/CRC, Boca Raton, FL, 2011.
7. E. Berchio, F. Gazzola, D. Pierotti, Gelfand type elliptic problems under Steklov boundary conditions, Ann. Inst. H. Poincaré Anal. Non Linéaire 27 (2010) 315–335.
8. M.G. Crandall, P.H. Rabinowitz, Bifurcation from simple eigenvalues, J. Funct. Anal. 8 (1971) 321–340.
9. S.I. Pokhozhaev, Eigenfunctions of the equation 1u + λf (u) = 0, Dokl. Akad. Nauk SSSR 165 (1965) 36–39.
10. H. Brezis, L. Nirenberg, Positive solutions of non-linear elliptic equations involving critical Sobolev exponents, Comm. Pure Appl. Math. 36 (4) (1983) 437–477.
11. Z. Guo, J. Wei, Global solution branch and Morse index estimates of a semilinear elliptic equation with super-critical exponential, Trans. Amer. Math. Soc. 363 (9) (2011) 4777–4799.
12. K.C. Chang, Methods in Nonlinear Analysis, Springer, 2005.

13. E.N. Dancer, A. Farina, On the classification of solutions of −1u = eu on RN : stability outside a compact set and applications, Proc. Amer. Math. Soc. 137 (4) (2009) 1333–1338.

14. [13] R. Bamón, I. Flores, M. del Pino, Ground states of semilinear elliptic equations: a geometric approach, Ann. Inst. H. Poincaré Anal. Non Linéaire 17 (5) (2000) 551–581.

15. J. Dolbeault, I. Flores, Geometry of phase space and solutions of semilinear elliptic equations in a ball, Trans. Amer. Math. Soc. 359 (9) (2007) 4073–4087.

16. I. Flores, A resonance phenomenon for ground states of an elliptic equation of Emden–Fowler type, J. Differential Equations 198 (1) (2004) 1–15.

17. I. Flores, Singular solutions of the Brezis–Nirenberg problem in a ball, Commun. Pure Appl. Anal. 8 (2) (2009) 673–682.

18. G. Arioli, F. Gazzola, Hans-Christoph Grunau, Entire solutions for a semilinear fourth order elliptic problem with exponential nonlinearity, J. Differential Equations 230 (2) (2006) 743–770.

19. G. Arioli, F. Gazzola, Hans-Christoph Grunau, Enzo Mitidieri, A semilinear fourth order elliptic problem with exponential nonlinearity, SIAM J. Math. Anal. 36 (4) (2005) 1226–1258.

20. J. Dávila, L. Dupaigne, I. Guerra, M. Montenegro, Stable solutions for the bilaplacian with exponential nonlinearity, SIAM J. Math. Anal. 39 (2) (2007) 565–592.

21. J. Dávila, I. Flores, I. Guerra, Multiplicity of solutions for a fourth order problem with exponential nonlinearity, J. Differential Equations 247 (11) (2009) 3136–3162.

22. J. Dávila, I. Flores, I. Guerra, Multiplicity of solutions for a fourth order equation with power-type nonlinearity, Math. Ann. 348 (1) (2010) 143–193.

23. G.R. Belickiˇı, N. Forms, Invariants and Local Mappings, Naukova Dumka, Kiev, 1979, p. 176.

24. E. Mitidieri, S.I. Pokhozhaev, A priori estimates and the absence of solutions of nonlinear partial differential equations and

inequalities, Tr. Mat. Inst. Steklova 234 (2001) 1–384; Proc. Steklov Inst. Math. 234 (3) (2001) 1–362. Translation.

25. D. Gilbarg, N.S. Trudinger, Elliptic Partial Differential Equations of Second Order, in: Classics in Mathematics, Springer-Verlag, Berlin, 2001.

26. E.A. Coddington, N. Levinson, Theory of Ordinary Differential Equations, McGraw-Hill Book Company, Inc., New York, Toronto, London, 1955.

27. J. Palis, W. de Melo, Geometric Theory of Dynamical Systems: An Introduction, Springer-Verlag, New York, Heidelberg, Berlin, 1982.

28. H. Brezis, J. Luis Vázquez, Blow-up solutions of some nonlinear elliptic problems, Rev. Mat. Univ. Complut. Madrid 10 (2) (1997) 444–469.

29. J. Dávila, L. Dupaigne, Comparison results for PDEs with a singular potential, Proc. Roy. Soc. Edinburgh Sect. A 133 (1) (2003) 61–83.

30. F. Faá di Bruno, Note sur une nouvelle formule de calcul differentiel, Quart. J. Pure Appl. Math. 1 (1857) 359–360

Index

R

S

T